Derivation and Use of Environmental Quality and Human Health Standards for Chemical Substances in Water and Soil

Other Titles from the Society of Environmental Toxicology and Chemistry (SETAC)

Veterinary Medicines in the Environment
Crane, Boxall, Barrett
2008

Relevance of Ambient Water Quality Criteria for Ephemeral and Effluent-dependent Watercourses of the Arid Western United States
Gensemer, Meyerhof, Ramage, Curley
2008

Extrapolation Practice for Ecotoxicological Effect Characterization of Chemicals
Solomon, Brock, de Zwart, Dyev, Posthumm, Richards, editors
2008

Environmental Life Cycle Costing
Hunkeler, Lichtenvort, Rebitzer, editors
2008

Valuation of Ecological Resources: Integration of Ecology and Socioeconomics in Environmental Decision Making
Stahl, Kapustka, Munns, Bruins, editors
2007

Genomics in Regulatory Ecotoxicology: Applications and Challenges
Ankley, Miracle, Perkins, Daston, editors
2007

Population-Level Ecological Risk Assessment
Barnthouse, Munns, Sorensen, editors
2007

Effects of Water Chemistry on Bioavailability and Toxicity of Waterborne Cadmium, Copper, Nickel, Lead, and Zinc on Freshwater Organisms
Meyer, Clearwater, Doser, Rogaczewski, Hansen
2007

Ecosystem Responses to Mercury Contamination: Indicators of Change
Harris, Krabbenhoft, Mason, Murray, Reash, Saltman, editors
2007

For information about SETAC publications, including SETAC's international journals, Environmental Toxicology and Chemistry and Integrated Environmental Assessment and Management, contact the SETAC Administratice Office nearest you:

SETAC Office
1010 North 12th Avenue
Pensacola, FL 32501-3367 USA
T 850 469 1500   F 850 469 9778
E setac@setac.org

SETAC Office
Avenue de la Toison d'Or 67
B-1060 Brussells, Belguim
T 32 2 772 72 81   F 32 2 770 53 86
E setac@setaceu.org

www.setac.org
Environmental Quality Through Science®

Derivation and Use of Environmental Quality and Human Health Standards for Chemical Substances in Water and Soil

Edited by
Mark Crane
Peter Matthiessen
Dawn Stretton Maycock
Graham Merrington
Paul Whitehouse

SETAC Technical Workshop
Faringdon, Oxfordshire, United Kingdom

Coordinating Editor of SETAC Books
Joseph W. Gorsuch
Gorsuch Environmental Management Services, Inc.
Webster, New York, USA

SETAC

CRC Press
Taylor & Francis Group
Boca Raton London New York

CRC Press is an imprint of the
Taylor & Francis Group, an **informa** business

CRC Press
Taylor & Francis Group
6000 Broken Sound Parkway NW, Suite 300
Boca Raton, FL 33487-2742

First issued in paperback 2019

© 2010 by Taylor & Francis Group, LLC
CRC Press is an imprint of Taylor & Francis Group, an Informa business

No claim to original U.S. Government works

ISBN-13: 978-1-4398-0344-8 (hbk)
ISBN-13: 978-0-367-38485-2 (pbk)

Library of Congress Cataloging-in-Publication Data

Derivation and use of environmental quality and human health standards
    for chemical substances in water and soil / editors, Mark Crane ... et al.
      p. cm.
    "A CRC title."
    Includes bibliographical references and index.
    ISBN 978-1-4398-0344-8 (hardcover : alk. paper)
    1. Water quality--Standards. 2. Soils--Quality--Standards. 3. Environmental
toxicology--Measurement. 4. Water quality biological assessment. 5.
Soils--Quality--Measurement. 6. Public health. I. Crane,
    Mark, 1962- II. Title.

TD370.D47 2010
363.6'1--dc22   2009032733

Visit the Taylor & Francis Web site at
http://www.taylorandfrancis.com

and the CRC Press Web site at
http://www.crcpress.com

SETAC Publications

Books published by the Society of Environmental Toxicology and Chemistry (SETAC) provide in-depth reviews and critical appraisals on scientific subjects relevant to understanding the impacts of chemicals and technology on the environment. The books explore topics reviewed and recommended by the Publications Advisory Council and approved by the SETAC North America, Latin America, or Asia/Pacific Board of Directors; the SETAC Europe Council; or the SETAC World Council for their importance, timeliness, and contribution to multidisciplinary approaches to solving environmental problems. The diversity and breadth of subjects covered in the series reflect the wide range of disciplines encompassed by environmental toxicology, environmental chemistry, hazard and risk assessment, and life-cycle assessment. SETAC books attempt to present the reader with authoritative coverage of the literature, as well as paradigms, methodologies, and controversies; research needs; and new developments specific to the featured topics. The books are generally peer reviewed for SETAC by acknowledged experts.

SETAC publications, which include Technical Issue Papers (TIPs), workshop summaries, a newsletter (*SETAC Globe*), and journals (*Environmental Toxicology and Chemistry* and *Integrated Environmental Assessment and Management*), are useful to environmental scientists in research, research management, chemical manufacturing and regulation, risk assessment, and education, as well as to students considering or preparing for careers in these areas. The publications provide information for keeping abreast of recent developments in familiar subject areas and for rapid introduction to principles and approaches in new subject areas.

SETAC recognizes and thanks the past coordinating editors of SETAC books:

A.S. Green, International Zinc Association,
Durham, North Carolina, USA

C.G. Ingersoll, Columbia Environmental Research Center,
US Geological Survey, Columbia, Missouri, USA

T.W. La Point, Institute of Applied Sciences,
University of North Texas, Denton, Texas, USA

B.T. Walton, US Environmental Protection Agency,
Research Triangle Park, North Carolina, USA

C.H. Ward, Department of Environmental Sciences and Engineering,
Rice University, Houston, Texas, USA

Contents

*Graham Merrington, Sandra Boekhold, Amparo Haro,
Katja Knauer, Kees Romijn, Norman Sawatsky, Ilse Schoeters,
Rick Stevens, and Frank Swartjes*

List of Figures

List of Tables

Acknowledgments

This book presents the proceedings of a Technical Workshop convened by the Society of Environmental Toxicology and Chemistry (SETAC) in Faringdon, Oxfordshire, United Kingdom in October 2006. The 35 scientists involved in this workshop represented 11 countries and offered expertise in ecology, ecotoxicology, environmental chemistry, environmental regulation, sociology, and risk assessment. Their goals were to examine the selection, derivation, and implementation of environmental quality standards.

The workshop was made possible by the generous support of many organizations, including

- AstraZeneca
- Department for the Environment, Food and Rural Affairs (United Kingdom)
- Environment Agency of England and Wales
- Dutch Ministry of Housing, Spatial Planning and the Environment
- Eurometaux
- Scottish Environment Protection Agency
- wca environment ltd.

We are also grateful to Professor Clive Thomson for expert and helpful peer review of the final draft chapters in this book.

About the Editors

Mark Crane is a director of wca environment (www.wca-environment .com). He has a first degree in ecology and a PhD in ecotoxicology and has worked on the effects of chemicals on wildlife for more than 20 years, in both consulting and academia. He has edited 3 books and published more than 100 papers on environmental toxicology and risk assessment. Recently, his work on environmental quality standards (EQSs) has included advice to industry  clients on water and biota standards under the Water Framework Directive, development of soil screening values and aquatic EQSs for the Environment Agency of England and Wales, and use of computational and read-across methods for deriving standards for the European Commission.

Dawn Maycock is a director of wca environment. She has a first degree in biology and environmental studies and a PhD in ecotoxicology. Recent work on EQSs has included using survival time analyses and species sensitivity distributions to derive time-specific EQSs for monitoring specific discharges, and development of aquatic EQSs for a range of substances under the Water Framework Directive (Annex VIII).

Graham Merrington is a director of wca environment. He has a first degree in environmental science and a PhD in environmental chemistry and has worked on the behavior and fate of metals in terrestrial systems for 15 years in academia, regulation, and consulting. He has published more than 50 papers on metals behavior and environmental chemistry. While working for the Environment Agency of England and Wales, he managed a research and devel-  opment program on EQSs focused on setting standards in soils, sediments, waters, and air. He has represented the United Kingdom at European Union Expert Groups for the Water Framework Directive and Existing Substance Regulations.

Peter Matthiessen is a consultant ecotoxicologist with nearly 40 years of experience researching the effects of pollutants on aquatic life. His main research interest since the 1980s has been endocrine disruption in fish, but he is also involved in the development of EQSs, in the environmental risk assessment of new chemicals, and in the development of improved methods for testing chemicals and monitoring their effects on the aquatic environment. He is a  member of the United Kingdom Advisory Committee on Pesticides, and cochair of the OECD (Organization for Economic Cooperation and Development) Validation Management Group for Ecotoxicity Tests.

Paul Whitehouse manages the Environment Agency's Chemicals Science team, based at Wallingford in Oxfordshire, United Kingdom. He has a first degree in botany from the University of Reading and a PhD in pesticide uptake and behavior in target species from the University of Bristol. Before joining the Environment Agency in 2004, he worked for 10 years as a researcher in agrochemicals with Shell Research Limited and subsequently as a  principal environmental toxicologist with the environmental consultancy WRc from 1991 to 2004. His main area of interest is the derivation and use of chemical standards for environmental protection and the associated underpinning science. He currently chairs the European Commission's Expert Group on EQSs, developing Europe-wide guidance for the derivation of EQSs under the Water Framework Directive.

Workshop Participants*

Marc Babut
CEMAGREF, Lyon, France

Graeme Batley
CSIRO, Bangor, Australia

Sandra Boekhold
Soil Protection Technical Committee, The Hague, The Netherlands

Mark Crane
wca environment limited, Faringdon, United Kingdom

Mark Douglas
Dow AgroSciences, Abingdon, United Kingdom

Andrew Farmer
Institute for European Environmental Policy, London, United Kingdom

John Fawell
wca environment limited, Faringdon, United Kingdom

Bernard Fisher
Environment Agency of England and Wales, Reading, United Kingdom

Maria-Amparo Haro
INIA, Madrid, Spain

Udo Hommen
Fraunhofer Institute, Schmallenberg, Germany

Thomas H Hutchinson
AstraZeneca, Brixham, United Kingdom

Martien Janssen
RIVM, Bilthoven, The Netherlands

Katja Knauer
Basel University, Basel, Switzerland

---

* Affiliations were current at the time of the workshop.

Chris Leake
Bayer Crop Science, Monheim, Germany

Robert Lee
University of Wales, Cardiff, United Kingdom

Stefania Loutseti
Dupont Agro, Athens, Greece

Peter Matthiessen
Independent consultant, Ulverston, United Kingdom

Steve Maund
Syngenta, Basel, Switzerland

Dawn Maycock
wca environment limited, Faringdon, United Kingdom

Graham Merrington
wca environment limited, Faringdon, United Kingdom

Paul Nathanail
University of Nottingham, Nottingham, United Kingdom

Adam Peters
Scottish Environment Protection Agency, East Kilbride, Scotland

Mary Reiley
USEPA, Washington, DC, United States

Kees Romijn
Bayer Crop Science, Monheim, Germany

Norman Sawatsky
Alberta Environment, Edmonton, Canada

Uwe Schneider
Environment Canada, Canada

Ilse Schoeters
European Copper Institute, Brussels, Belgium

Kieron Stanley
Environment Agency of England and Wales, Bristol, United Kingdom

Rick Stevens
USEPA, Washington, DC, United States

Bill Stubblefield
Parametrix, Corvallis, Washington, United States

Frank Swartjes
RIVM, Bilthoven, The Netherlands

Jacqui Warinton
Syngenta, Bracknell, United Kingdom

Tony Warn
Environment Agency of England and Wales, Bristol, United Kingdom

Lennart Weltje
BASF AG, Limburgerhof, Germany

Paul Whitehouse
Environment Agency of England and Wales, Wallingford, United Kingdom

1 Introduction

Mark Crane, Martien Janssen, Peter Matthiessen,
Steve Maund, Graham Merrington, and
Paul Whitehouse

1.1 BACKGROUND

Chemical standards are widely used to protect the environment and human health from substances released by human activity. Generally, standards relate to doses or concentrations in the environment for specific chemicals, below which unacceptable effects are not expected to occur. Many standards are legally enforceable numerical limits, such as Environmental Quality Standards for List 1 chemicals in water or Annex X and VIII standards under the European Water Framework Directive. Others are not mandatory but are contained in guidelines, codes of practice, or sets of criteria for deciding individual cases. Some standards are not set by governments but carry authority for other reasons, especially the scientific eminence or market power of those who set them (e.g., World Health Organization guidelines).

The ways in which Environmental Quality and Human Health Standards are derived, and the frameworks within which they are used, differ between countries and regions, with standards derived, expressed, monitored, and implemented in different ways. To some extent, this diversity reflects genuine technical differences that must be taken into account in the development of standards for different compartments (e.g., water or soil) or for different receptors (e.g., humans, livestock, or flora and fauna). However, much standard setting has been developed in a piecemeal fashion with little consistency between schemes in the levels of protection sought, the selection of chemicals for which standards may be needed, the methods used to derive them, or the methods used to monitor compliance. These differences can lead to the implementation of substantially different values from the same empirical data, which must mean that their application is either over- or underprotective in at least some situations.

The SETAC Technical Workshop on the Derivation and Use of Environmental Quality and Human Health Standards for Chemical Substances in Water and Soil was held in Faringdon, Oxfordshire, United Kingdom, from October 16 to 19, 2006. The workshop addressed the methods by which substances are selected and prioritized for standards derivation, the way in which these standards are derived (e.g., EU

Technical Guidance Document and Water Framework Directive approaches; EU member state, North American, and other international approaches) and the way in which they are implemented (e.g., mandatory pass or fail; probabilistic, e.g., 95th percentiles; or tiered risk assessment frameworks). Soil and water standards were considered, as were values for the protection of human health and the natural environment. The focus was on European regulatory frameworks, although expert input was sought from other jurisdictions internationally. Chemical standards for aquatic (water and sediment) and terrestrial (soil and groundwater) systems were the main focus for the meeting. This workshop built on, and included some participants from, a 1998 SETAC workshop *Re-evaluation of the State of the Science for Water-Quality Criteria Development* (Reiley et al. 2003).

1.2 WORKSHOP OBJECTIVES AND TOPICS

The workshop brought together 35 scientists and professionals from 11 countries. These individuals had expertise across risk assessment, environmental and health sciences, and social science and economics disciplines; they included toxicologists, chemists, risk assessors, economists, sociologists, and managers, as well as regulators and policy makers. These individuals were selected to provide sectoral (academic, government, and private sector), geographical, and gender balance to the workshop.

The proposed objectives and topics of the workshop were as follows:

1) Scientific and risk assessment of aquatic (water and sediments) and terrestrial (soil and groundwater) data for derivation of standards
   a. How should substances be selected and prioritized for standard setting?
   b. Which biological assessment endpoints should be used for setting standards (e.g., community, population, individual, cellular, biomarker, 'omics, etc.)?
   c. How should data be assessed for reliability and relevance?
   d. How should background or ambient concentrations of substances be considered when deriving or using standards?
   e. How can standards be validated or verified, and what should be the role of field (e.g., epidemiological) and semifield (e.g., microcosm or mesocosm) data?
   f. What environmental and human health risks from substances should we protect against and at what level of protection?
   g. How should uncertainty be taken into account when setting standards?
   h. What is the appropriate choice of assessment factors and statistical extrapolation models in relation to data quantity and quality?
   i. How should regional, national, and site-specific risks be taken into account?
   j. What are the differences in the scientific and risk assessment approaches used to derive environmental versus human health standards, and can these differences be justified?
   k. What are the differences in the scientific and risk assessment approaches used to derive aquatic versus terrestrial standards, and can these differences be justified?

2) Implementation analysis and assessment of technological options
   a. At what point should implementation analysis be performed, and what should be the analysis inputs and outputs?
   b. What techniques should be used for implementation analysis?
   c. What are the most effective implementation strategies?
   d. How should compliance monitoring statistics and practice be defined for standards?
   e. At what point should technological options be assessed when deriving a standard, and what should be the assessment inputs and outputs?
   f. What assessment techniques should be used to assess technological options?
   g. Can standards be used to drive technological innovation?
   h. What are the differences in the implementation analysis and assessment of technological options for environmental versus human health standards, and can these differences be justified?
   i. What are the differences in the implementation analysis and assessment of technological options for aquatic versus terrestrial standards, and can these differences be justified?
3) Social and economic appraisal of standards
   a. At what point should the costs and benefits of particular standards be appraised, and what should be the appraisal inputs and outputs?
   b. When is a cost-effectiveness analysis sufficient?
   c. What techniques should be used in cost-effectiveness and cost-benefit analyses?
   d. At what point should the public and direct stakeholders be involved in the derivation and use of standards?
   e. What are the advantages and disadvantages of public and stakeholder involvement?
   f. What are the best involvement techniques and strategies?
   g. What is the appropriate use of "expert judgment" when deriving or using standards?
   h. What are the differences in the socioeconomic analysis of environmental versus human health standards, and can these differences be justified?
   i. What are the differences in the socioeconomic analysis of aquatic versus terrestrial standards, and can these differences be justified?

Workshop participants were assigned to one of four breakout groups:

1) Scientific and risk assessment approaches for derivation of environmental and human health standards for the aquatic environment (water and sediment)
2) Scientific and risk assessment approaches for derivation of environmental and human health standards for the terrestrial environment (soil and groundwater)

3) Technological appraisal and implementation of environmental and human health standards for aquatic and terrestrial environments
4) Socioeconomic analysis of environmental and human health standards for aquatic and terrestrial environments

The results of the workshop are contained in this monograph. Chapter 2 outlines the social and economic frameworks within which standards are derived and used and provides a wider context for considering when, where, and how standards should be implemented. Chapter 3 continues this theme and explores general issues associated with implementing any standard. Chapters 4 and 5 consider the detail of deriving and implementing aquatic and terrestrial standards. Finally, in Chapter 6 we draw together the overall conclusions of the workshop and provide recommendations on the development and use of standards. We also identify future research that would help to underpin the science of environmental and human health standards.

REFERENCES

Reiley MC, Stubblefield WA, Adams WJ, Di Toro DM, Hodson PV, Erickson RJ, Keating FJ Jr. 2003. Re-evaluation of the state of the science for water-quality criteria development. Pensacola (FL): SETAC Press.

2 Setting Environmental Standards within a Socioeconomic Context

Andrew Farmer, Robert Lee,
Stefania Loutseti, Kieron Stanley,
Jacqui Warinton, and Paul Whitehouse

2.1 INTRODUCTION

This chapter is concerned primarily with the technical activities involved in setting environmental standards. However, standard setting is not an entirely technical pursuit. Those responsible for recommending new standards are increasingly expected to look at the costs of introducing a new standard and to regulate in a way that retains public trust. Regulators and scientists must take account of these realities.

Clearly, we must develop standards that are sensitive to the social and economic context so that we can be confident that a new standard strikes a sensible balance between the environmental benefits that arise from the standard and the costs and implications of meeting it.

2.2 SOCIAL AND ECONOMIC CONTEXT

2.2.1 ECONOMIC ANALYSIS

The aim of any regulation can be seen as a way of curing some event of market failure, that is, to intervene in the market because it is failing somehow to direct resources to their most valued use. The usual consequence of market failure in the case of environmental pollution is when social and environmental costs (e.g., degraded water resources denying people access to clean drinking water) are not reflected in the price of goods on the market. If a polluting industry is allowed to make free use of environmental media, to the disadvantage of those dependent on or deriving utility from those media, then the cost of environmental damage is not reflected in the goods placed on the market. This is commonly referred to as an "externality." Regulation looks to internalize the cost of environmental damage, for instance by recognizing that cost and making the polluter pay. It is important that a balance is drawn between the costs of pollution abatement and the improvement in the environment as a result of abatement activity.

Setting standards at a level that is too strict or lenient may mean that the socially optimal level of pollution control is not met. In practice, the investment needed to meet the standard could exceed the value of any benefits. Alternatively, proposals for such standards may simply be rejected, in which case the effort expended has been wasted, and there are no effective controls on the potentially polluting chemical substance. We therefore need to ensure that the process for deriving standards strikes a fair balance between the need to protect the environment and human health without imposing excessive burdens on industry, restricting legitimate ways of earning a living, or producing unnecessary work for regulatory agencies that have to implement the standard.

2.2.2 SOCIAL ASPECTS OF STANDARD SETTING

We often hear that trust in the decisions about human health and environment made by government has rarely been lower, yet public expectations about transparency and consultation are growing. The development of environmental standards is no exception, but levels of public and stakeholder engagement in the process are currently low, and even where they take place at all, usually it is only after proposals are well developed.

We argue that those involved in standard setting need to respond to these concerns. Specifically, we need to ensure that the process for delivering standards takes account of any unintended consequences (e.g., the costs of meeting a standard), and that the outcome provides a reasonable balance between environmental protection and regulatory costs. Furthermore, the process must be transparent and auditable. This means it should be participatory, communicative, and inclusive. If this can be achieved, the outcomes are more likely to be trusted by all those who are affected by the standards that are eventually introduced.

It is helpful to make clear some principles and assumptions that continue throughout the rest of this chapter. We first examine the different types of standards that are used (typological issues) and their role in environmental regulation. We then outline a framework for deriving a new standard that attempts to incorporate social and economic factors into our decision making.

2.3 TYPOLOGICAL ISSUES IN UNDERSTANDING STANDARDS

2.3.1 INTRODUCTION

It is important to recognize that there are different types of standards; Table 2.1 provides some examples of different terms that have been used. This list does not include the technical terms that are sometimes used within that process (such as predicted no effect concentration, PNEC), but those that tend to be used in public documents.

TABLE 2.1

Examples of terms used for "standard"

Terms	Comment
Standard	Viewed as a number that should be reached. It may or may not be mandatory.
Goal	An aspirational objective, usually long-term, guiding action, but not mandatory.
Limit (or limit value)	A level that should not be breached. In EU law, a limit value is legally binding.
Guide value	A level that should be aimed for, but is not binding.
Indicative value	A level that is taken as an indicator of quality but is not binding.
Benchmark	Some form of nonbinding value, often to form the basis for comparative assessments of environmental status, such as between locations or over time.
Trigger value	A value that should lead to further action. This might be a legal obligation or a requirement to understand better the risks to wildlife or humans. The action could be to provide public information.
Threshold	A nonbinding value indicating a change in status.
Criteria	Values that can be used to guide further action.

Numerical standards may be further expressed in different ways (e.g., as percentiles, averages, or absolute maxima). These different expressions of a standard are dealt with in more detail in this chapter.

2.3.2 Why Does Terminology Matter?

The public largely uses the term "standard" in everyday discourse as an indication of what they might be entitled to expect. In contrast, policy makers and regulators require a range of different types of standards in their environmental management decision making, allowing regulatory flexibility in the way that the value is to be used. Over time, several terms have been developed, including those shown in Table 2.1.

It is important, however, to note that the terms used by policy makers (and in law) are not consistent. For example, under EU law binding air quality standards are termed "limit values." However, the Integrated Pollution Prevention and Control (IPPC) Directive, which regulates industrial pollution, uses the term "environmental quality standard" and makes clear that the limit values are environmental quality standards. Both technical experts and lawyers, therefore, can use different terms in similar contexts. This variation not only can be found between media (e.g., air and water standards) but also can occur within a single medium.

Equally, the same term can be used in different contexts. In the United Kingdom, air quality standards are adopted by government, but these are not legally binding. However, legally binding values referred to as standards can be adopted under EU law (e.g., the Dangerous Substances Directive).

Clearly, this variation in the terminology is a potential problem in public engagement. We do not propose to set out new definitions but simply to highlight this difficulty in the language used to increase the awareness of those involved in standard setting.

2.3.3   WHY ARE DIFFERENT TYPES OF STANDARDS NECESSARY?

2.3.3.1   Standards for Different Purposes

It is entirely appropriate to have different types of standards, but the consequences of using the different types illustrated in Table 2.1 vary. For example, a prescriptive, legally binding standard can have limited flexibility and requires potentially costly action to be taken if it is breached (in other words, if the level in the environment exceeds that stipulated in the standard). Therefore, it is critical that the scientific analysis undertaken to develop such a standard is as robust as possible, and in particular, areas of uncertainty in the process need to be carefully assessed and communicated. A standard that is intended as a guide for providing information can take into account the same issues as a legally binding standard, but uncertainty could be treated in a different way because the implications of passing or failing the standard are quite different.

We must be clear about what the standard is meant to do. Each standard must also be accompanied by clear statements of how the standard was derived, such as the range of factors taken into account (the receptors it is intended to protect and costs of compliance) and how uncertainty is treated. This is necessary as it is not unknown for standards developed for one purpose to be used in other contexts — sometimes inappropriately. Such clarity will help stakeholders decide what a reasonable application for a standard is and when a particular application might be inappropriate.

2.3.3.2   Selecting the Right Standard for the Job

It is vital also to consider the regulatory framework in which a standard is to be set because this largely determines what type of standard is required.

Those seeking to constrain environmental externalities will have a number of options at their disposal. At one extreme, goods may be considered safe to circulate freely on the market, with no prior approval and with little warning or labeling regarding use. On the other hand, substances may be so hazardous that they need to be banned from sale on the market. Between these two positions there is a range of other environmental management options (Figure 2.1). These options go beyond traditional direct regulation to include indirect economic methods, using incentives or taxes to change behaviors.

Command-and-control processes are the traditional way in which environmental pollution has been regulated. In typical form, this will amount to a license to emit a chemical or waste product to the environment but with conditions attached to the license that limit the quantity that can be released. Discharge consents to water are an example of this type of measure.

However, the environmental objective might best be addressed by seeking to change behavior through an advisory standard (e.g., one that acts as an information

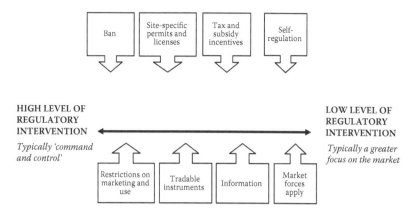

FIGURE 2.1   Possible mechanisms for controlling chemical risks to the environment.

source rather than some sort of legally binding limit). But, we might be aware of significant behavioral problems with a particular industry sector, indicating that a more prescriptive approach is desirable. It might become apparent that there is insufficient knowledge (environmental, toxicological, economic, etc.) to develop a binding standard (even if this is desirable), in which case an alternative type of standard is the only feasible option. Such issues would need to be clearly identified early in the process because they influence what the final standard can look like.

In comparison to command-and-control standards, those built into taxes or other market mechanisms have certain advantages. They allow industry to choose when to invest in pollution abatement while providing a longer-term incentive to make that investment. The amount of investment will be determined by the levels of the tax, but a carefully weighted tax on emissions will provide a framework to ensure that spending on abatement does not exceed that which is socially optimal. Because market actors determine their response in the light of the incentives, regulatory discretion is minimized. On the other hand, command-and-control structures offer greater certainty of outcome in terms of a fixed emissions limit for any facility, providing the system is well policed. Differentiated standards can respond to local conditions (e.g., discharge permits that take account of variations in available dilution), whereas a tax is likely to apply in a strict manner that is less sensitive to such differences in local conditions.

Although we tend to focus on command-and-control systems and economic incentives as the main instruments of regulation, Figure 2.1 shows there is a wide range of other contexts in which a standard might be used. What is important is to realize that almost all measures taken to abate environmental pollution through regulation will require some sort of standard in devising effective regulatory models.

Text Box A explains each of these different approaches to regulation — and the role played by standards — in more detail. Our main points are that

- It is appropriate to have different types of standards.
- There are many more ways to set and implement a standard than legally binding limits that are introduced through "direct" regulation. However,

even the "softer" approaches entail the use of a standard that must have a sound basis in science.

- Knowledge constraints might limit the type of standard in some cases.
- The basis for a standard needs to be clearly described so that standards are not used inappropriately.
- There should be flexibility in the way a standard is applied to reflect the levels of uncertainty associated with the standard and the consequences of failure.

TEXT BOX A: DIFFERENT APPROACHES TO ENVIRONMENTAL REGULATION — THE ROLE OF STANDARDS

COMMAND AND CONTROL

Command-and-control processes include licenses to emit a chemical or waste product to the environment (e.g., discharge consents to water). However, there is flexibility here in the way that a failure to meet the consent condition is handled. Exceeding the consent condition constitutes an administrative or criminal offense, but persistent offending may lead to withdrawal of the license to operate. The precise nature of these mechanisms can vary. On the one hand, liability may be strict with no available defense once the illegal discharge is proven. Alternatively, there may be a negligence-based standard under which either the regulator has to prove absence of care or the polluter has an available defense to show that, in spite of the excess discharge, reasonable care was taken. We might therefore wish to use a less-stringent standard if liability is strict, but a more stringent standard if reasonable care affords a good defense.

Over time, regulatory systems have tended to combine emission limits with other license requirements set in terms of the pollution control technology to be employed by the licensee. An example would be the requirement for best available techniques (BATs) in Europe under the IPPC Directive and best available technologies in the United States under the Clean Air and Clean Water Acts. These systems have built-in technology forcing that seeks to drive forward standards by innovation and investment in technology, with potential savings for the industry (such as waste reduction) or economic opportunities for those providing environmental goods and services.

Command-and-control regulation may simply require some form of technology to abate environmental impacts, such as wastewater treatment, or it may combine this with a level to which the technology must work. Although this can be based on an ambient concentration, the standard will usually be translated into a limit at the point of discharge, taking account of local site-specific conditions (e.g., available dilution). We often base our judgment of compliance with a permit on sampling and chemical analysis. This introduces issues about sampling and analysis errors and the level of confidence we need before declaring that a permit has been breached. This important technical aspect is dealt with in Chapter 3.

Finally, standards may be applied in this way across an industrial sector (uniform emission limits) or may be differentiated with different burdens reflecting different levels of risk to the environment or different prospects of abatement.

INCENTIVES FOR CHANGING BEHAVIORS

In recent years, there has been some move away from command-and-control systems toward incentives like those illustrated in Figure 2.1. Charges imposed on the polluter would be one example. Classically, private law might do this by awarding damages against an operator that has caused environmental damage. Although this creates some difficulty in relation to the unowned environment (since law tends to compensate those with an interest in land), both the United States and the European Union have developed liability systems to internalize costs resulting from environmental damage. Such awards are after the event, however, and we may wish to ensure some form of ex ante incentive to prevent damage at the source.

One of the measures shown in Figure 2.1 is an environmental tax. For example, if we decide that pesticide use is causing diffuse pollution to groundwater, then we might tax the purchase of pesticides in an attempt to encourage management practices with lower pesticide use. Setting of the tax level is vitally important; a farmer might be expected to reduce harm caused by pesticides until the point at which the costs of alternative production begin to outweigh the tax. Under these circumstances, we might choose to subsidize rather than tax (i.e., to pay those farmers who forgo pesticide use where environmental improvement would result). All such price mechanisms will incorporate some consideration of environmental quality standards because an acceptable level is implied, even if we do not go out to measure compliance. Effectively, we are treating the standard as a design target.

A rather more subtle instrument might be a tradable allowance, a mechanism that harnesses the allocation of rights to pollute described under command-and-control systems. If an industry is permitted to discharge into the environment up to an assigned level but actually discharges below that limit, then it is allowed to trade the excess capacity with those operators likely to exceed their limits. This type of mechanism has been employed in Europe as the favored route to curbing greenhouse gases. Again, environmental standards are necessary because they form the basis of trading. They need to be set in such a way that the total allowance is environmentally beneficial (curbing greenhouse gases) while providing sufficient incentives in the system (in terms of the value of the tradable instruments) to encourage active trading.

2.4 FRAMEWORK FOR DERIVING A NEW STANDARD

The preceding discussion explained that standard setting needs to take into account a variety of technical, economic, policy, legal, and social factors. Furthermore, there can be a good deal of flexibility in the way a standard may be used. We argue that

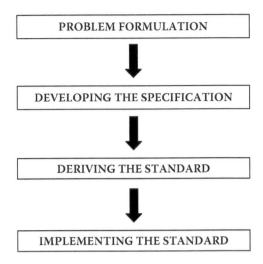

FIGURE 2.2 Key steps in the development of an environmental standard.

an effective approach to standard setting needs to address these issues before we embark on a scientific analysis of the data. For this purpose, a framework is helpful to guide the activities that are needed and the sequence in which they are handled.

Our proposed framework (Figure 2.2) draws on the recommendations of the Royal Commission on Environmental Pollution (RCEP) in their 21st report (1998) but attempts to take this further by offering practical steps to take forward the principles highlighted by the RCEP. For the purposes of this review, we focus on numerical standards that describe concentrations, loads, or doses of chemicals in the ambient environment. We also highlight where costs and benefits should be considered and opportunities for useful interaction with stakeholders and the general public. Details of the technical activities that would be undertaken in deriving and implementing the standard are covered in other sections of this chapter.

Although we do not advocate a strictly linear approach, it is most easily communicated through a series of steps, each comprising simple checklists that the risk assessor and manager might usefully adopt.

2.4.1 PROBLEM FORMULATION

Like most complex technical activities, a clear understanding of the problem that needs to be addressed will shape the work that follows and reduce the risk of wasted effort. We cannot assume that the problem formulation for a standard is self-explanatory. For example, which chemicals are priorities for standard development? What are the regulatory or social drivers that require us to reduce risks to the environment? Is a new standard actually the best way of achieving our environmental goals? Are there any legal or political constraints on the approach (e.g., those illustrated in Figure 2.1) we can adopt? Does past experience lead us to favor one approach over another?

Problem formulation will logically continue into the specification stage. The specification makes operational the issues identified in the problem formulation; whereas problem formulation sets out "what" needs to be done, the specification sets out "how" this is to be achieved.

2.4.1.1 Why Is a Standard Needed?

By identifying the drivers for a new standard, the rationale for the work can be more readily communicated to those who might be affected. Table 2.2 sets out examples of possible drivers behind the "need" for a standard.

Once the key drivers have been identified, it is helpful to consider what, if anything, would change if we did nothing before scoping the specification for a standard. This is in effect a zero-based review in which the need for a standard is set against the possibility of taking no action at all. If it is clear that some risk reduction is necessary, this does not necessarily mean that we must develop a standard, but it reinforces the argument for taking some sort of action. If there is a clear legal requirement to set a standard, then this step would be obvious.

2.4.1.2 Who Needs to Be Involved?

The aim of this stage is to identify who can help with, or might be affected by, a new standard. We must also identify people with the appropriate skills to develop and deliver the standard. This step is important in helping promote transparency in the process. If trust can be cultivated at this early stage, it is also more likely to help unlock sources of data that might otherwise be withheld.

We would expect that the key stakeholders are given the opportunity for engagement throughout the process, up to the time a standard is implemented. However, the level of effort expended should reflect the importance of the standard. For substances of only local or minor concern, it could be as simple as defining potential

---

TABLE 2.2

Possible drivers for setting a standard

Major incident/accident	Incidents involving substances with the potential for unexpected adverse effects or response to accidents involving such substances
Implementation outcome	Often the legacy of existing standards, such as previous time-limited standards or introduction of standards for new areas
Regulatory/statutory trigger	National or international law requiring a new standard to be set
International conventions	International law (treaty obligations) requiring compliance with standards, such as OSPAR (Oslo and Paris Commissions)
Emerging risks	Possible changes in chemical use, or new chemicals found, that suggest a risk to health or the environment that requires control
Political pressure/public concern	Presence of a chemical in the environment no longer socially acceptable even if it is not known that it will have adverse effects
Third-sector pressure	Lobbying from nongovernmental organizations, think tanks, and charities

work packages with locally based colleagues. For substances that occur widely, would affect many industrial activities, or are known to be hazardous, we may need to prepare for public and industry engagement. Text Box B gives an example of good practice in public engagement. Identification of stakeholders is a useful starting point. Some standard-setting bodies have existing guidelines on participation (e.g., UNESCO [United Nations Educational, Scientific, and Cultural Organization]), and for statutory activities, guidelines on consultation may already exist.

Typical stakeholders might include

- The policy or business user of the standard
- Technical specialists (e.g., toxicologists and epidemiologists)
- Social and economic researchers
- Holders of data in industry or industry organizations (e.g., on toxicity or chemical monitoring in the environment)
- Statutory consultees who must legally be included

TEXT BOX B: GOOD PRACTICE IN PUBLIC ENGAGEMENT — THE FEDERAL ADVISORY COMMITTEE ACT IN THE UNITED STATES

BACKGROUND TO FACA

In 1972, the Federal Advisory Committee Act (FACA; Public Law 92-463, 5 U.S.C., App.) was passed by Congress. Its purpose was to ensure that advice given to government by the "various advisory committees, task forces, boards, and commissions formed over the years by Congress and the president, be both objective and accessible to the public."

In 1996, the US Environmental Protection Agency (USEPA) set up a federal advisory committee composed of members with interests ranging from the environmental and agricultural communities to state and local governments. The committee's objective was to recommend ways to improve the effectiveness and efficiency of state, territorial, tribal, and USEPA total maximum daily load (TMDL*) programs.

The committee's report was issued in July 1998 (USEPA 1998). It contains recommendations based on broad agreements reached by the members of the federal advisory committee.

Key recommendations are as follows:

- Restoring impaired waters must be a high priority.
- Implementing TMDLs is key to a program's success.

---

* TMDL: a calculation of the maximum amount of a pollutant that a water body can receive and still meet water quality standards and an allocation of that amount to the pollutant's sources.

> - Communications with the public are critical.
> - Stakeholder involvement is key to successful implementation.
> - The government's capacity to evaluate TMDLs must be strengthened.
> - An iterative approach to TMDL development and implementation is the best way to make progress in uncertain situations.
>
> The proposed regulatory changes were published in the *Federal Register* for public review and comment before the legislation was agreed.

2.4.1.3 Constraints

Experience shows that a program of standard setting is almost always subject to some sort of constraint. If constraints are identified at the outset, then time and money are not wasted on formulating proposals that could never be implemented. Such constraints might include

- *Resources available.*
- *Legal constraints.*
- *Scale*, as in the case of air standards requiring coordination across borders (global, regional, or local).
- *Timetable.* There may be a legally imposed timetable for introduction of new standards that could limit the amount of analysis that can be undertaken.
- *Methodology.* Some statutory regimes stipulate the method that must be used to derive new standards.
- *Unwillingness* by some partners to participate (e.g., in providing data).
- *Data applicability and availability.* There may be a need to collect further data, or gather nonexpert evidence. For example, the limited transferability of data between freshwaters and marine waters could be a constraint. There is a need to ensure that all data are fit for the intended purpose.
- *Legacy of assumptions.* There may be relevant existing standards, but they may include scientific and nonscientific assumptions that need to be revised in the light of new understanding.
- *Permissible typologies.* There needs to be a clear view on what types of standard are permissible, especially whether the standard should be expressed as a legally enforceable limit, a trigger for action, or an aspirational target. This will often depend on the regulatory regime in which the standard is to be used.
- *Expectations of the regulated community.* What will people expect from the standard and access to the decision-making and knowledge-gathering processes?

As we will explain, it may be prudent to consider different options for a standard for the same substance, or even different typologies. This can help assign the uncertainties involved in deriving the standard to the environment or producer or to share them. In other words, we can adopt a precautionary interpretation of the available data and give the benefit of doubt to the environment. Alternatively, a less-precautionary interpretation would accord the benefit of doubt to those emitting the chemical. Economic efficiency might dictate that we need some flexibility in the level of environmental protection afforded by the standard. For example, a high level of protection might be disproportionately expensive to achieve. In this case, a range of options can be helpful so that we can assess the environmental protection afforded by each option against the costs of achieving it. If such flexibility is permissible, it must be highlighted at this stage because it will place extra demands on the scientists analyzing the data. If flexibility is introduced later, it will understandably appear to be a change based on expediency, and the standard could be subjected to understandable criticism or even lose credibility entirely.

Although we do not anticipate wide engagement with nonexperts on every standard, there are some issues for which policy makers might want better understanding of public values about what they expect from the environment, for example, on the level of environmental protection afforded by chemical regulation. Manufacturers may have valuable knowledge of how substances are used in practice or about complaints concerning exposure to these substances. Approaches to capture nonexpert knowledge can be done in a variety of ways and need not necessarily include expensive techniques such as citizens' juries or facilitated events. Simple events, such as workshops where groups can be invited and given an opportunity to discuss the implications of a standard, are a means of avoiding confrontation while gathering potentially valuable insights from a broad spectrum of potentially affected communities. Informal discussions with specific communities of interest can also be a cost-effective way of anticipating concerns.

In addition, guide or indicative values might very well be used to open broader public scrutiny and examine in more detail regional variations without burdening the standards regime with the full economic costs of communicating and consulting widely on binding standards. This would be made easier if forums have already been put in place to engage with the public.

2.4.1.4 Social and Economic Questions

Sometimes there are minimum requirements on those charged with setting standards to consult and assess the geographic dimensions of the standard. A standard could benefit or disadvantage specific communities with particular economic interests (e.g., fishing communities). Those involved in standard setting will need to be sensitive to these concerns as well as public values and to consider whether these might affect the standard-setting process or the way the standard is implemented. The expected level of consultation is usefully logged in the problem formulation stage but should be updated before the specification is finalized.

The full economic costs of upstream engagement and its value to the standard-setting process can be difficult to quantify. However, the staggering costs incurred by the UK government in introducing genetically modified organism (GMO) field trial, highlight the opportunity costs (costs incurred due to the failure to act) of not conducting adequate upstream and downstream engagement.

The problem formulation step should therefore consider which steps are needed to sustain an equitable and open process and to make appropriate provision for this. This stakeholder analysis would normally include two aspects:

1) What types of engagement can be considered? Surveys, focus groups and citizens' juries are all well-established methods for engaging with stakeholders (Wilsdon et al. 2005). However, we must ensure that the costs are proportionate to the problem. Workshops and visits to key stakeholders may be more appropriate in some circumstances.
2) Should public values influence levels of spending and control? There is little point in gathering stakeholder opinion if these opinions will not inform the specification in any way or the levels of activity in developing a standard. Moreover, the development of a specification may critically depend on the activities of stakeholders (such as in providing data to develop the standard) or understanding who is responsible for enforcement. Clarification on these points will prevent problems later in the process.

2.4.1.5 Stakeholder Analyses

It is helpful to understand whether stakeholders represent "communities of interest" or "communities of place." A community of interest may be a body of individuals, institutions, industries, and so on with an interest in an issue (e.g., charities devoted to clean air or anglers lobbying for clean water). A community of place will be a community or set of communities inhabiting an area that will be affected by a particular standard (e.g., a riparian community).

Both of these groups have much to contribute to the process but from different perspectives. Again, the opportunity costs of not engaging as widely as possible might include spiraling economic costs associated with poor implementation of a standard. They might also lead to differences in the success with which a standard is implemented between locations.

We argue that an economic appraisal should form a key part of the development of a standard, on a par with technical assessments of toxicological and environmental data. Achieving a balance between these aspects is most readily addressed by early engagement to gauge the appetite for being included and to facilitate engagement. The level of such engagement is likely to be higher when there is more at stake, such as when the standard is to be a statutory measure applied widely across many activities, affecting many people or businesses. When problem formulation indicates that an aspirational or trigger value is appropriate, a much lower level of engagement with communities of interest or place — and possibly none — may be appropriate.

2.4.1.6 Check Rejection Criteria

A check on rejection criteria avoids developing a standard that is not fit for purpose, and includes a reasoned assessment of the costs relative to the benefits of the proposed standard. Although this will be a preliminary assessment, it will be important in helping focus attention on those aspects that could prevent a workable standard from being developed.

In short, stakeholders (especially policy makers) must agree with the following:

- There is a need for a standard.
- The technical ability exists to deliver it and monitor it in a fit-for-purpose manner.
- There are no insurmountable constraints.
- The stakeholders have been identified.
- There is a mandate agreed from which the process can begin.

If these criteria cannot be met, it will be necessary to return to earlier steps in the problem formulation process.

2.4.2 DEVELOPING A SPECIFICATION

Developing the specification is a critical part of the standard-setting process. It follows logically from the problem formulation stage. The specification makes the following clear:

- The preferred type of standard (including any types that are not acceptable)
- Advice on implementation criteria (in the case of a standard that will be subject to compliance assessment)
- How the standard should be expressed
- The geographical scope (where it will apply)
- The technical methodology, legal context, and nature of socioeconomic assessments needed

Preparing and implementing a "stakeholder engagement plan" may help achieve more comprehensive engagement, especially if we can expect a lot of sensitivity about a new standard. Consideration might be given to ways of publicizing the intention to develop a standard and how the specification was derived. Any conflicts between stakeholders should be addressed through communication in a fair and transparent manner.

A draft should be made available to stakeholders who might reasonably expect to be consulted or their advice sought (e.g., on basic assumptions, data, risks and frequency of exposure, and ways in which key uncertainties might be reduced). For example, industry may have access to toxicity or higher-tier biological data that can reduce an important area of uncertainty (and thereby allow a smaller assessment factor to be used or even permit a different extrapolation approach). This makes the standard-setting process more participatory, more interactive, and less likely to be challenged later. It can make the process slower, at least at the beginning, but if data become available through these interactions, we may avoid the need to revisit aspects of the standard-setting process later. The development of a standard may well be an iterative process, revisiting the specification, development, and implementation stages if new constraints become apparent (e.g., scientific data are lacking) or the economic implications are different from what was anticipated.

The following steps describe the key elements for a specification.

2.4.2.1   Scope

The scope of the standard must be documented. To do this effectively, it is important that all parties involved should be clear about the aims of the proposed standard. In addition, clear definitions (e.g., Table 2.1) are important to avoid ambiguity and to ensure that there is shared understanding of the objective. It will be necessary to define any activities or situations that should be specifically excluded from the standard and to provide guidance on when the standard would be applied. Documenting reasons for including or excluding some situations will be useful when decisions need to be communicated or justified. Where the standard is required to meet legal obligations, there may be no flexibility.

2.4.2.2   Form of the Standard

Standards may be expressed in a number of forms (e.g., as concentration, mass, or load/unit area). The specification must be explicit but should identify options where there is genuine flexibility. The use of the standard will dictate whether it can be expressed as an absolute limit or some sort of time-weighted percentile.

Vitally, policy makers need to consider the implementation of the standard at this early stage, in particular the degree of flexibility in the way it may be expressed and implemented. Chapter 3 deals with implementation issues in detail, but standards can fail to be implemented because these factors were not addressed early enough in the process.

For a standard that is to be used within a formal compliance assessment regime, we must obviously define the limit value (the "magnitude") of the standard. However, the scientific analysis will also need to address four other criteria that are needed for a defensible standard. These are dealt with in detail in Chapter 3; in summary, these are as follows:

- Summary statistic (e.g., met for 95% of time): frequency
- Period of time for calculation (e.g., a calendar year): duration
- Definition of design risk (acceptable degree of failure, e.g., 1 year in 20)
- Statistical confidence in demonstration of compliance (e.g., 95% confidence that failure occurs for more than 5% of the time)

Decisions about each of these criteria are determined by 1) the substance and its pattern of use and 2) by the degree of rigor with which we want to apply the standard:

1) The expression of a standard will depend on the use of a substance and its underlying distribution of occurrence in the environment. For example, a standard expressed as the average concentration over a year may be entirely appropriate for a contaminant that occurs widely and throughout the year, while a high percentile may be a more sensible approach when the chemical occurs only sporadically and for short periods. Many pesticides would

fall into this latter category (e.g., contamination by spray drift from a short-lived but potent insecticide). For such chemicals, consideration of the permissible return period might be an important criterion.

2) Normally, regulators adopt a default value of 95% confidence that the standard has been breached for mandatory standards. However, there are circumstances under which this might be changed. The level of confidence (i.e., the burden of proof) could be lower (e.g., 90% or even 50%) if we are seriously concerned about the presence of a particular contaminant or if there is a good reason to adopt a particularly cautious approach (e.g., a release to a designated nature reserve). Conversely, we might be prepared to take action only if we were very confident that the standard had been exceeded. This might apply if there is high uncertainty in the derivation of the standard (such as a lack of data for certain taxonomic groups), it is very costly to take remedial action, it is only an "interim" standard, or the standard applies to emissions to a heavily degraded environment.

There are other ways in which we might introduce flexibility in the way the standard is implemented. One is to allow a transitional period during which a discharger should work toward achieving the standard but need not comply. This might be done by delaying the onset of the regulation to allow investment in process technology or investigation of options for mitigation. Alternatively, we might change the status of a standard to an "advisory" limit or guideline, perhaps with advice on best practice to users and dischargers about how it might be achieved but then moving to a mandatory compliance limit over time. Of course, a final option is to retain its mandatory status but to relax the limit value to one that can be achieved without excessive costs. However, we must be aware that such an approach could compromise health or environmental protection, at least in the short term. There are many instances of staging targets, for example, to increase the conservation status of sites of special scientific interest by increasing the percentage that can meet this target over time.

Our point is that there are various ways in which flexibility in the rigor of a standard may be introduced, and it is vital that these are recognized at the outset. Just how much flexibility is allowable and how that may be introduced are policy decisions. The specification should be as explicit as possible about this. In practice, the scientific analysis will influence what options are permissible, but we must at least guide the direction and scope of the scientific analysis.

2.4.2.3 Monitoring

Where a standard is to be subjected to monitoring, the specification would normally highlight certain operating conditions that will need to be satisfied. They include

- *Reference conditions:* These are physical or chemical parameters that need to be assessed in conjunction with the analysis for standard compliance (e.g., temperature, pH, organic carbon content for soil or dissolved organic carbon, and total suspended solids for surface water) (depending on which factors influence chemical behavior and toxicity).

- *Analytical sensitivity:* Analytical methods need to be adequate so that compliance with the standard can be quantified. The specification should not detail the method, but it needs to raise the expectation that a description of the analytical method and its performance characteristics (limit of detection ideally better than 5 to 10 times below the lowest regulatory level and maximum precision and bias targets at the regulatory limit) will form part of the standard. Data quality requirements are also important to ensure that reliable conclusions may be drawn from analytical measurements (e.g., the total error of measurement should not exceed a certain percentage at a given level of sensitivity, such as 30% of the value of the standard). For empirical methods such as leaching tests or assessing bioavailability of a given parameter, for which the method defines the result, a prescribed method should be cited.

  Several methods may be available for a particular analyte. Limiting these to a few options may help achieve consistency, particularly over a larger region, and contribute to accuracy, reliability, and, thereby, fairness. The options for monitoring should be as wide as possible as long as they can deliver a sufficient level of accuracy and precision. This will allow leading-edge technology to be used as soon as it becomes available and not 3 to 10 years later when the relevant international standard incorporating this new technology becomes available.

- *Sampling for compliance assessment:* As explained, the form and expression of the standard (e.g., annual average, 95th percentile) will largely dictate the sampling regime that is required. The specification should draw attention to the need for guidance on sampling frequency, the need for spot or composite samples, and any pretreatment of samples (e.g., filtration or stabilization) prior to analysis. The importance of employing suitable sampling and sample preservation techniques must be stressed.

2.4.2.4 Consideration of Costs and Benefits

A rigorous assessment of the costs and benefits of setting a standard will be necessary for standards that are to be legally binding and could have serious cost implications. However, standards that are advisory or provisional will require less detailed social and economic analysis. There is certainly more leeway in examining the costs to industry of complying with a standard that it is not obliged to meet.

Therefore, the specification should identify how the costs and benefits of a standard are to be assessed and the approach to be used. At its most basic level, it will include an appraisal of the extent of failure for a particular standard and an approximation of the financial cost of reducing that failure. Clearly, contamination by some chemicals will be more difficult and costly to mitigate than others.

There are two main approaches to informing decisions that take account of economic factors. One is cost benefit assessment (CBA) and the other is multicriteria decision analysis (MCDA). It is not the purpose of this document to provide detailed guidance about these techniques, but we offer some comments in the context of standard setting (Text Box C). For a more detailed appreciation of CBA, refer to

---

TEXT BOX C: MCDA OR CBA?

Multicriteria decision analysis helps individuals make a decision (in our case, an environmental standard) from a series of options. Essentially, it considers criteria that will be important to a decision (which might be environmental protection, costs of compliance, impacts on particular livelihoods), uses value-scaling techniques to assign weights to each of the criteria according to their importance, and takes account of the likelihood of each one occurring. CBA, on the other hand, is invariably concerned with economics, facilitating a decision that takes account of the combined costs and benefits of a possible option to all parties. Although both have an important role in decision analysis, there are also some important differences.

In the context of standard setting, MCDA seems to be a more appealing approach than CBA. This is because

1) MCDA allows a wider range of noneconomic factors (e.g., species diversity, difficulties in sampling or analysis, public values about the environment, impacts on different industry sectors) to be explicitly factored in rather than simply "considered," as they are in CBA.
2) The health or environmental benefits (e.g., reduced incidence of disease, protection of particular species, or biological diversity) do not need to be monetized.
3) It is possible to terminate an MCDA early; we can use it simply to place options before us in an unbiased way, leaving the final decision to policy makers (thus it operates as a multicriteria analysis, MCA).
4) The value functions and criteria weightings used in MCDA can be solicited from different stakeholders in an open and transparent way. The conversion of a performance measure to a monetary value in CBA can be rather opaque.

However, CBA may be suitable if the protection goal concerns an economically important entity, such as Atlantic salmon populations or forestry products.

The Green Book: Appraisal and Evaluation in Central Government (2003); and for MCDA, to Dodgson et al. (2000).

2.4.3 Deriving a Standard

The technical aspects of developing a standard are covered in Chapters 4 and 5. Our aim is to generate a standard that provides a considered balance between the benefits of using the standard and the costs of achieving it. We argue that a scientifically robust standard that cannot be implemented is as ineffective as a standard that is scientifically flawed. It is also important that we are able to justify our decisions, and this will be aided by open and transparent deliberations.

2.4.3.1   Integrating Scientific, Social, and Economic Factors

If the consequences of setting a standard will have important environmental, economic, or social consequences, we envisage an approach in which a social and economic analysis forms part of the overall decision in conjunction with the scientific analysis. The two activities come together in an MCDA that seeks to integrate all the factors that will deliver the required level of environmental protection at an acceptable economic and social cost. If that cost can be defined during problem formulation, this will greatly facilitate the process because it puts limits on what is, and is not, permissible.

In its early stages, the process might show that a particular standard simply cannot be achieved, for example, because there are too few data, it is too expensive to implement or monitor, or it becomes clear that a certain typology is likely to be ineffective. This could force a return to the initial specification for the standard.

Assuming we are able to proceed, we need to understand the following:

1) The relationship between the magnitude of a standard and biological impact
2) The relationship between other implementation aspects of a standard (e.g., acceptable degree of failure and statistical confidence in demonstration of compliance) and biological impact
3) The cost of mitigating exposure, including an understanding of who bears those costs (e.g., particular industry sectors, households, or regulatory agencies)
4) Other unintended consequences (e.g., added complexities in sampling and analysis)

By combining this information in an MCDA we can estimate the cost and other implications of achieving different levels of environmental protection and balance these in coming to a final decision (Figure 2.3). The value of effective stakeholder engagement becomes apparent here, especially if controls on a substance are likely to be borne more heavily by some sectors than others. An example might be hill farmers, whose use of chemicals to control external parasites in sheep is important to their livelihoods. These groups are likely to ascribe different weightings to the costs of mitigating exposure than others. In this case, there might also be an animal welfare criterion informing the final decision.

2.4.3.2   Understanding the Relationship between Exposure and Effects

A key element in decision making will be to understand the relationship between the level of chemical exposure and the consequent risks to health or the environment. There are two main ways in which we can understand this relationship: through the species sensitivity distribution (SSD) or the dose–response curve. The SSD is perhaps the more useful for environmental assessment because it integrates all species, whereas the dose response describes the cause–effect relationship for only one species. Nevertheless, the dose–response relationship could be a valuable tool for environmental assessment when the species described is either particularly sensitive,

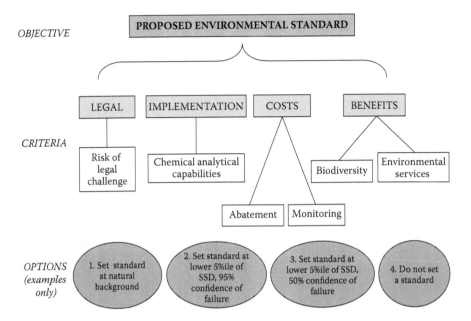

FIGURE 2.3 Multicriteria decision analysis for assessing options for an environmental standard.

of economic value (e.g., a prized game fish), or is highly valued for some other (e.g., conservation) reason. In MCDA, this would be a critical input in informing the relationship between mitigation options and the consequent environmental benefits.

Also, MCDA allows the costs and practicality of meeting a standard to be accommodated in the final decision. This can be achieved by identifying the technological options for mitigating exposure, each of which would be associated with a different standard. They could include not only a "do nothing" option but also the application of different technologies or assumptions about the benefits that would follow from adopting "best practice" in some or all industry sectors. This may require stakeholder input to help focus attention on the most feasible abatement options. A preliminary analysis may usefully be shared with stakeholders so that they have an opportunity to comment and provide further information (e.g., to refine assumptions or prompt data collection to reduce uncertainty).

If data permit, it would be helpful to estimate margins of safety associated with different courses of action. Using MCDA, we could then compare the impacts associated with the estimated exposure that result from adopting a particular abatement technology with other abatement options or even no abatement at all. This allows policy decisions to be made in an informed manner. There are examples of this in areas such as fish stock preservation, and although it might be necessary to accept that political or economic considerations will ultimately determine standards, at the very least decisions are made in an informed and transparent manner. If contaminant concentrations are likely to give rise to serious impacts even after introducing a range of abatement steps, we might need to revisit the policy goal and specification.

The analysis need not be confined to the estimation of a standard that uses conventional methods of extrapolation (e.g., default assessment factors). It might usefully be accompanied by an analysis of the consequences of exceeding a particular standard. If a standard is proposed that sits within the range of available toxicological or ecotoxicological data or close to the lower limit of experimental data, then we should try to predict what impact an exceedance will have (e.g., excess cancer risk, proportion of species affected, or number of individuals affected).

As well as costs of compliance, there may also be economic "benefits" to the business of environmental improvement. For example, removal of contaminants at source from a relatively small volume of effluent saves water companies money because they do not have to invest in removing them from very much larger volumes of drinking water. Similarly, tourism benefits from a cleaner environment (e.g., compliance with bathing water standards, restoration of angling fisheries, or promotion of water sports). There may also be important health benefits that can be monetized, such as changes in deaths brought forward or working days lost. As far as possible, these also should be estimated and allowed for in the MCDA.

For ecosystems, there is a concept of "environmental services" (e.g., soil function). Soil conservation provides a range of obvious benefits, including the preservation of land use and fertility, supporting the production of fibers, food, or biofuels. These services have widened over time to encapsulate benefits of carbon sequestration or biodiversity protection. It is relatively easy to place values on the productive capacity in the former examples. It is somewhat harder to find the value of the latter uses, but as the Stern report (Stern 2006) illustrated, it is by no means impossible to calculate the value of, say, an upland peat bog.

There are sometimes genuinely different interpretations of the same data (e.g., the significance of a particular biological effect or the impact that pH has on the toxicity of a substance). An added virtue of the MCDA approach, especially if it can be done collectively, is that all criteria enter the evaluation, which should help build some degree of consensus among those holding different views. Nevertheless, it might be necessary to recognize genuine differences in interpretation and permit a range of values to emerge from the process rather than a single value. If there are insurmountable differences between the values, we cannot recommend a legally binding standard and must instead default to something such as an advisory goal, or a stepwise approach, moving to more stringent conditions over time.

2.4.4   IMPLEMENTATION OF STANDARDS

Unless a standard is actually implemented, its development would remain a purely academic exercise. The final step in the proposed framework deals with implementation of standards. However, different types of standards have very different implementation requirements, ranging from comprehensive sampling and compliance assessment schemes within some sort of legal regime to the agreement of text for a warning label.

For a legally binding standard, implementation is relatively clear (as long as the standard itself is clear). Monitoring of a medium to which the standard applies will indicate whether implementation has, or has not, been achieved. Similarly, if there are guide or indicative values, implementation can be assessed in the same way, although there may be no obligation to achieve those values (e.g., one might simply report against them). If standards are advisory, then implementation issues may become less stringent, but monitoring remains important.

A first critical issue in examining implementation is to ensure that a standard is clear. The numerical values must be expressed in terms that are understood (magnitude, period, etc.) and cover all the criteria outlined, even if this is only to explain that they are not relevant in a particular situation. The standard should also seek to be consistent with others that are in use (as far as is possible), so that a new standard is an extension of already accepted decision-making processes. Implementation also requires a standard to be practicable. If a standard limit value is set below background (natural concentrations), it is clearly impractical. If a standard is set but too little time is given to manage discharges of the pollutant, then this might also be impractical.

2.4.4.1   Meeting a Standard

For the purposes of this discussion, it is assumed that to meet a standard (in any medium) regulators would impose controls on activities (e.g., discharges to water) of the substance that could lead to a breach of that standard. These controls might limit an activity (e.g., application of nitrogen to land), a process (e.g., a limit on the concentration or mass of a substance that may be emitted), or a requirement to utilize proven abatement technology. Discharge limits will usually be based on a generic limit value, so they share a number of similar elements to standards (e.g., they should state magnitude, period of application, etc.). Some might view an emission limit value as a "standard" (indeed, it is a type of performance standard), and the public often conflate discharge and environmental quality standards when asking what standards "apply" to an industrial activity. Our point here is that standards that apply in the receiving environment are often translated into site-specific permits; compliance may be assessed against these as well as, or instead of, the standard applied to the receiving environment.

2.4.4.2   Allowing Flexibility

Flexibility can be important in setting a standard. There are various reasons why this might be the case:

- The standard applies to a wide geographic area (e.g., the European Union), so that local environmental conditions could still be identified that affect its implementation.
- The business environment might be under rapid change, so that the cost implications of the standard could change and assumptions in the standard-setting process become invalid.
- There might be sufficient scientific uncertainty in the technical analysis of a standard to mean that a rigid numerical outcome is inappropriate (note that while one option could be not to develop a standard if uncertainty is high, this might not be politically acceptable).
- Social concerns might vary over space and time and will need to be taken into account.

We have already introduced some ways in which flexibility can be introduced into standards by adopting different types or, when a compliance target is used, by addressing the criteria outlined. If flexibility is too broad, there is a risk of bringing too little pressure to bear, and environmental improvements are seriously compromised. Other sources of flexibility include the following:

- Setting a standard as a range, reflecting the intended use of a water body or land area, or the starting position (protecting pristine waters and improving poor waters could use different standards)
- Allowing the standard to be exceeded for particular periods of time (e.g., number of days per year)
- Requiring action if a standard is exceeded (e.g., closure of a bathing water beach or redesignation)
- Allowing those affected by a standard some period of opting out to give time for implementation (e.g., use of "variances" by individual states in the United States or use of derogations under the EU Water Framework Directive)

Whatever implementation measures are adopted, they will need to be explained so the approach can be justified and the decision made transparent. If the standard-setting process has been open and participatory but the implementation is opaque, then credibility is lost (to the regulator, industry, etc.), and trust is reduced (Farmer 2007).

We must also be prepared to enforce a standard if it is important enough. It is inappropriate to use such measures to achieve what is, in effect, a permanent reduction in a standard. For example, if exceedances of a standard routinely occur and are simply accepted, then what the public might think is that compliance with an annual average concentration of $10 \, mg \, L^{-1}$ is in fact only compliance with a concentration of $20 \, mg \, L^{-1}$. If the latter is environmentally acceptable, then it should have been used as a standard in the first place. Only if they are steps on a path toward full compliance should such interim measures be tolerated (e.g., for particular "hot spots" or problem periods).

2.4.4.3   Taking Socioeconomic Factors into Account

If an environmental quality standard is not met, a regulator would not normally be expected to engage in assessing stakeholder views or to analyze the benefits and costs of action in achieving that standard.

It is likely that there might be different sources of the substance for which a standard has been set, each requiring different approaches to control each of those sources. Thus, there is a need to consider the optimum distribution of costs in an equitable way — those responsible for the greatest impact should bear the larger part of the cost. The deadline for compliance will also provide further opportunities for economic analysis, examining different investment strategies, for example.

Many control measures affect not only the substance for which a standard has been set but also other contaminants as well. For example, improved water treatment may reduce emissions of other contaminants. Therefore, the analysis undertaken for implementation should take this into account in examining benefits and costs.

2.4.4.4   Importance of Feedback

When there are implementation failures, it is important to record these and engage stakeholders in understanding why they have occurred and what the consequences might be. This will vary depending on the nature of the standard. For example, if a standard is based on extrapolation using large assessment factors, then exceedances of the standard by a factor of just 2 or 3 might mean little risk to biota, or there is the possibility of a greater risk to some species, but these cannot be identified. However, if assessment factors are small, then we can be more confident about the consequences. Such statements assist in creating a regime of trust.

2.5   SOME FINAL THOUGHTS

Standards are not purely scientific constructs but also are important regulatory tools that need to be set at a level that strikes an appropriate balance between the level of protection they afford and the investment needed to meet the standard (i.e., to achieve a socially optimal level of pollution control). We argue that this optimization should be an integral part of the standard-setting process. Experience suggests that the "process" for deriving standards is more likely to be challenged than the "values" themselves. Therefore, a value that might seem arbitrary to the public is more likely to be accepted if it can be shown to have emerged from a trusted regime that is clear and auditable. This challenges us to be able to show how an outcome was reached and what assumptions were made and to be open about the remaining uncertainties about which we can do nothing.

With these principles in mind, we propose a framework for deriving environmental standards that addresses not only the necessary technical considerations but also the social and economic ones. In particular, establishing an early relationship with stakeholders should make it easier to discuss openly any issues about data requirements or uncertainties identified later in the development of a standard. Within this framework, we argue for a formal approach to identifying a preferred option that optimizes the balance between potentially conflicting scientific, social, and economic factors, especially if the proposals could be contentious or expensive. MCDA is a tool that seems to address this balancing task, in which technical, economic, and social factors all contribute to a final decision.

Throughout, we have tried to identify where flexibility in a standard may legitimately be introduced. There are many different types of standard, and they all have a place in controlling chemical exposure in the environment. However, the type of standard to be used, and any measures to assess compliance with it, must be decided at an early stage. This is a policy decision. As well as the type of standard, flexibility can be introduced in the magnitude of the standard itself (it may be more or less protective) and, for standards that are subject to formal compliance assessment, in the way we decide whether the standard has been passed or failed. There is clearly more to a standard than merely the concentration, dose, or load of a substance. Aspects such as design risk, return period, and confidence of failure are integral features of the standard and, as such, should be subject to the same scrutiny.

REFERENCES

Dodgson J, Spackman M, Pearman A. 2000. Multi-criteria decision analysis: a manual. London: Department of the Environment, Transport, and the Regions. Available from: http://www.communities.gov.uk/pub/252/MulticriteriaanalysismanualPDF1380Kb_id1142252.pdf..

Farmer AM. 2007. Handbook of environmental protection and enforcement. London: Earthscan.

[RCEP] Royal Commission on Environmental Pollution. 1998. Setting environmental standards. 21st report. London: HMSO.

Stern N. 2006 The Stern review: the economics of climate change. London: Cabinet Office – HM Treasury.

The green book: appraisal and evaluation in central government. 2003. London: Stationery Office.

[USEPA]. US Environmental Protection Agency. 1998. Report of the Federal Advisory Committee on the Total Maximum Daily Load (TMDL) Program. The National Advisory Council for Environmental Policy and Technology (NACEPT). United States Environmental Protection Agency, Office of the Administrator (1601F) EPA 100-R-98-006, July 1998, Washington, DC, USA.

Wilsdon J, Wynne B, Stilgoe J. 2005. The public value of science: or how to ensure that science really matters. London: Demos.

3 How Should an Environmental Standard Be Implemented?

Mark Crane, Bernard Fisher, Chris Leake,
Paul Nathanail, Adam Peters, Bill Stubblefield,
and Tony Warn

3.1 INTRODUCTION

Environmental quality standards (EQSs) are widely used to help protect the environment and human health and are considered by many to consist simply of the stated "limit value" for a substance plus, perhaps, the time over which the standard applies. An example of this might be an annual average EQS for lead in water of 7.2 µg L^{-1} to protect the freshwater environment as proposed in the Water Framework Daughter Directive on EQS (European Commission [EC] 2006).

This chapter shows that other information must also be considered before an environmental standard can be implemented successfully. The implementation of standards cannot be a totally science-based issue; technical, social, and economic factors must also be considered. Critically, as discussed in Chapter 2, the legal or policy context must be clear from the start, and a standard based on scientific knowledge should then be applied to a specific policy context. However, in most situations there are few data and an incomplete understanding, which leads to uncertainty in the standard itself and, potentially, to its application. This uncertainty must be accounted for if a standard is to be applied consistently and fairly (Royal Commission on Environmental Pollution [RCEP] 1998).

In this chapter, we consider in more depth the following issues:

- Types of standards
- Essential implementation features for a standard

3.2 TYPES AND USES OF STANDARDS

The term "standard" can mean different things to different people. It is also used for different purposes (Chapter 2, Table 2.1). Standards can be

- The "number" produced by a toxicology or ecotoxicology study (e.g., an LD50 or a no-observed-effect concentration [NOEC]).

- Conservative (i.e., stringent and precautionary) in nature and with no regulatory enforceability, such as the Canadian Water Quality Standards and Eco-Soil Screening Levels.[1]
- Guidelines within a process of decision making that may use other information to corroborate success or failure. The outcome might be to seek more information or to impose controls (e.g., use of predicted no effect concentrations [PNECs] in Existing Substances Regulations in the EU).
- Intervention values or "the concentration above which land might present 'unacceptable' risk" (Department for Environment, Food and Rural Affairs [DEFRA and EA] 2002; Nathanail 2005).
- Legally enforceable limits when compliance is mandatory.
- Used in relation to pollution incidents to assess their impacts.
- Used to limit the discharge of chemicals from point sources.

Standards can be expressed in various units, such as a load (mass/unit area), a dose (mass/body mass), or a concentration (mass/volume; mass contaminant/mass soil). The use of a unit depends on the environmental compartment under consideration. We might consider the amount of pollution in water (a concentration), the consequences of equilibrium uptake by (or exposure to) a human (a dose), or the acceptable uptake by an ecosystem under steady conditions (the loading). The choice of unit depends on the point in a cycle or pollutant linkage at which we set the standard, as illustrated in Figure 3.1.

For convenience, we might set the standard at the point where the material enters the environment, as an "average" (or modeled average) concentration within the compartment, or as a value within a species. For example, for nitrogen in agriculture the standard is set in terms of load. However, the critical load depends on the compartment, such as heathland or grass, and whether trying to protect the vegetation from changing because of excessive nutrients or favoring species that prefer acidic conditions. The standard is a convenient way of setting a level of protection but is a simplification for dealing with a complex situation. The responses of individuals or ecosystems to pollution are likely to be very variable. Setting standards in terms of

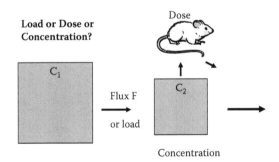

FIGURE 3.1   Transfer of potentially toxic substance between environmental compartments C_1 and C_2.

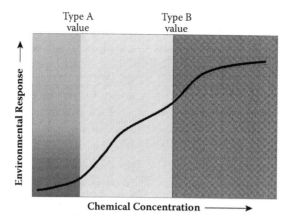

FIGURE 3.2 Response function superimposed on screening and mandatory standards.

simple numerical values can only be a guide to whether harm is likely to occur and is generally associated with considerable uncertainty.

For effective implementation of standards, users must be provided with an understanding of the strengths and limitations of the method of deriving the standard, and they must be helped to appreciate the type and quality of the underpinning data. This will help them decide if the standard is appropriate for its intended use and how to deal with small exceedances of only a few percent.

Current practices result in a range of different approaches and numerical values, depending on the jurisdictional procedures, the data used in the standard-setting process, and the intended use of the value. Inherent in this is that the degree of conservatism or precaution that is "embedded" in the standard needs to be clearly understood and reported.

By and large, it is possible to place standards in, or between, 2 groups, as shown in Figure 3.2. Here, it is assumed that as the concentration of a chemical in the environment increases, an environmental response is increasingly likely to occur; the response — or probability of a response — will increase with concentration.

Figure 3.2 illustrates that environmental standards can be set in at least 2 ways:

1) Type A values are standards that "if met are not anticipated to result in adverse environmental or human health effects." Exceedence of these standards does not necessarily lead to adverse environmental consequences because such standards tend by design to be conservative and precautionary. Examples of these are values such as the US Environmental Protection Agency's (USEPA's) ambient water quality criteria (USEPA 2000) or the UK soil guideline values (DEFRA and EA 2002).

- Type B values are standards that "if exceeded are anticipated to result in adverse environmental effects." An example of Type B values might be the standards used to take action after pollution incidents or standards for which failure leads directly to a regulatory response. Examples include the mandatory standards in European directives on bathing waters, freshwater fish, and the now-withdrawn Interdepartmental Committee on the Redevelopment of Contaminated Land (ICRCL) action levels (ICRCL 1987; Nathanail 2005)

It is of course important for a particular standard to know where it lies in the spectrum from Type A to Type B. The range of values from better than Type A to worse than Type B spans the range of probability of unacceptable environmental consequences.

Policy to protect pristine water or land from future development might use Type A standards. Criteria for deciding when to take expensive measures to clean up polluted water or land with existing development might be based on standards more like Type B. In the former, the benefit of doubt in setting the standards is given to the environment or the exposed human population. In the latter, it is shared or given to the polluter. Once the first stage of cleanup has worked, a next stage might focus more on moving toward the Type A standard.

There is an echo of this typology in the use of classification systems, such as those that apply under the EU Water Framework Directive (WFD). High status under the WFD might use Type A standards. The classification may then pass through good, moderate, poor, and bad status, where poor or bad might be defined in terms of failure of Type B standards. A similar idea is used in the guideline (Type A) and imperative (more like Type B values) standards in several European directives on water quality (e.g., directives for bathing waters, shellfish water, freshwater fish, and surface water for abstraction for drinking water; see also Section 4.1).

A complementary way to classify standards (Table 3.1) is to consider how uncertainty over the "true" value of the standard might be related to whether it is used as a screening tool (where uncertainty is high and a precautionary approach to the uncertainty has been taken) or as a legally mandatory pass or fail threshold (for which there is little doubt) (Figure 3.3).

TABLE 3.1

Examples of meaning of screening and mandatory standards

Screening standards	Mandatory standards
Minimal consequences	Significant consequences
Lower tier of assessment	Higher tier of assessment
No unacceptable effects	High probability of effects
Examples	
Intermittent discharges	Dangerous Substances Directive standards (UK)
Sediment standards (UK)	Nitrate (UK)
	Bathing water (UK)
Threshold effect levels (TELs) (US)	Probable effect levels (PELs) (US)

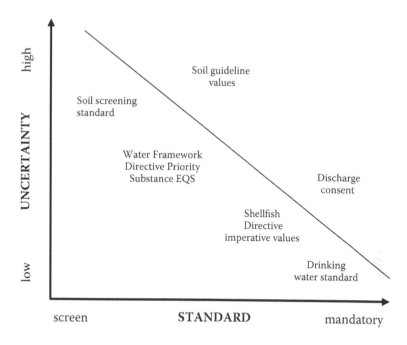

FIGURE 3.3   Representation of uncertainty over spectrum of standards.

Within all of this, there is the additional and separate issue of the statistical confidence of failure or compliance that is associated with taking decisions. For any standard, monitoring (or modeling) can produce outcomes in the range from 100% confidence of compliance to 100% confidence of failure. If the consequences of failure are controversial or expensive, it makes sense to demand that failure is demonstrated with high confidence before such action is imposed on the unwilling. This gives the statistical benefit of the doubt to the polluter. If less-dramatic action is to be imposed (e.g., a requirement for further studies), we might act on less confidence of failure. Costly action must not be imposed solely because a regulator has chosen not to monitor adequately.

Similarly, if it is important to be reassured that there is no unacceptable risk of damage, a regulator might demand that compliance is demonstrated with high confidence. This gives the benefit of the statistical doubt to the receptor. Both these requirements for confidence, and the degree of failure or compliance that can be detected, should feed into the design of compliance assessment (sampling) programs.

There is a trade-off between these considerations of confidence of failure and use of the Type A or Type B standards illustrated by Figure 3.2. A Type A standard and a demand that compliance is demonstrated with high confidence is doubly precautionary: The benefit of doubt in derivation of the standard and the benefit of doubt from monitoring are both assigned to the receptor.

3.3 ESSENTIAL FEATURES THAT ALLOW IMPLEMENTATION OF A STANDARD

3.3.1 USING STANDARDS TO GUIDE DECISION MAKING

3.3.1.1 Absolute Limits versus Ideal Standards

Standards might be used to respond to pollution incidents, for example, to demonstrate that a particular event killed a group of fish 30 minutes ago. Such a standard might be used in legal action against the culprit. In this case, the standard is likely to be used to protect against acutely toxic events, such as a pesticide spill or air pollution episode or an occasional discharge, which may occur for a short period of time. Such a value could be expressed as a high percentile (e.g., 99.9th percentile), although for practical reasons it may be more appropriate to regulate such instances on the basis of an absolute limit. Monitoring may, for example, only be undertaken during or shortly after an event, thus preventing any statistical assessment of the monitoring frequency. Such standards may be required when there is a need to identify possible acutely toxic episodes that may occur irregularly or infrequently, although this may simply be for the practical reasons of implementing and enforcing such standards.

Similarly, absolute limits are popular with lawyers; guilt or innocence is (assuming no sampling or analysis issues) more clear-cut and simple to explain.

However, it is also common to use standards to set up the infrastructure, policies, controls, or rules that mean that incidents and risks occur with acceptably rare probabilities. These standards might be described as "strategic" standards. For example, controls on ammonia in sewage treatment works (which are back-calculated from environmental standards) are designed to promote good fisheries in the receiving river. The intention is to reduce serious incidents to an acceptable frequency in each river because the infrastructure of sewage treatment appears to function at this level of acceptable risk. This may result in a compromise, which is essentially that standards are set up as particular types of summary statistics and not as absolute limits.

In summary, those recommending toxic limits as standards should provide advice on where their "number" sits in the range from [guideline, trigger] to [mandatory, imperative], as shown in Figure 3.2.

The spectrum of standards between guideline or intervention values and mandatory or action values should also be aligned with the uncertainty surrounding the standard and the consequences of failure. Standards that have a high degree of uncertainty surrounding them should be implemented as guidelines or trigger values, and the consequences of failure should not be excessively severe. In such cases, failure of the standard may require that a more detailed investigation and risk assessment is carried out (e.g., use of soil guideline values in land contamination risk assessment; cf. DEFRA and EA 2002). Standards that have a high degree of certainty surrounding both the toxicological criteria and the assessment of compliance should or could be implemented as mandatory or imperative standards with significant consequences of failure (e.g., emission controls or marketing and use restrictions).

Those recommending standards should also state whether their "number" is self-contained. Is failure of their standard all that is needed to justify decisions? Otherwise, they should describe how the number is to be used with other information in the decision-making process, specifying what this other information might be and how it should be used. For example, in sediment quality assessment, some jurisdictions require that a sediment quality value (i.e., a chemical quality standard) is used within a "triad" approach in association with information from sediment bioassays and benthic ecological survey results.

The difference between decision making on the basis of a standard alone and on the basis of a standard plus other information can be described as one aspect of the direct versus the indirect model. In the direct model, action is defined exactly as that needed to secure compliance with the standard, for example, as permit conditions for discharges to water that are calculated to meet an environmental standard in a river. The classic cases here are substances like ammonia or cadmium in rivers and other mandatory standards in various European directives.

In the indirect model, action may be only a "step in the right direction," perhaps with little certainty that this will achieve compliance. This might involve general controls on operators in an area showing failure of an environmental standard (e.g., controls on agriculture as a response to failure to meet nitrate standards in rivers or groundwater). This kind of control is becoming more common because we face more and more issues for which knowledge of cause and effect is unclear. The principle is that at some point in the future another step in the right direction might be necessary.

Also, those recommending standards need to explain how failure should be defined and determined and so help estimate the potential extent of failure in the environment: That is, what percentage of water bodies fail a particular standard? This is required, in principle, even if there is little hope that money will be found to make estimates of actual compliance. It is a first step toward knowing who is causing failure and so to an estimate of the costs and benefits to society of using the standard.

The subtext here is that some types of standards will not be used by regulatory authorities if failure cannot be estimated reliably.

3.3.1.2   "Ideal Standard"

All of the information discussed requires that each component of a standard is defined in the manner of an "ideal standard" (RCEP 1998). This is a standard for which values are defined in 5 specific areas (as described in this section). An ideal standard opens the way for the use of mitigation measures that are triggered in a consistent way from site to site, that meet consistent endpoints, and that allow an unbiased assessment of compliance before and after remediation. They can also be used to produce unbiased comparisons of sites, regions, and nations.

Five points that robust standards should reflect are set out here as they would apply to a chemical in freshwater where compliance is assessed by routine periodic monitoring. They are developed also for wider circumstances, in particular for cases such as contaminated land, for which spatial rather than temporal variability is a prime concern. The 5 points are that the standard includes

1) A measurable limit value (e.g., a concentration of 10 μg L^{-1}).
2) A summary statistic, such as how often the limit may be exceeded (e.g., 5% of the time or on 5% of the monitoring events). This point excludes the absolute limit because compliance and planning for such standards can only be

done by defining the absolute limit as a particular percentile. Instead of an absolute limit, statistics such as the mean or percentiles are required.
3) The period of time over which this statistic is calculated, such as a calendar year.
4) The definition of the design risk, which is the proportion of time periods for which failure to meet the standard is considered to be acceptable, such as 1 in 20 calendar years. This will be relevant in designing the action needed to correct failure. In other words, when this action is complete, how often is it acceptable for a failure to be reported?
5) The statistical confidence with which noncompliance is to be demonstrated before remedial action is taken in response to a failure.

Points 1 to 3 explicitly define the assessment of compliance. Point 1 is what many scientists would regard as "the standard," but it is only a part of the standard. Points 2 and 3 deal with the requirement that the limit value needs to be used for 2 purposes: to estimate the measures needed to correct or prevent failure and to assess compliance in an unbiased manner (perhaps in a way that enables comparisons between regions or nations). Both of these tasks require standards defined as summary statistics that can be estimated using statistical methods in an unbiased manner, thus allowing the quantification of statistical errors. Generally, this means that the standards must be expressed as summary statistics such as annual averages and annual percentiles.

Point 4 is a consideration mainly in the design of schemes to achieve compliance. It considers what the risk of failure should be when the remedial action is complete.

Finally, given Points 2 and 3, we can deal with Point 5, which defines how certain we wish to be that exceedance of a standard has occurred. There is an aspect of confidence that is concerned with the reliability of the standard in the first place and the evidence used to create it. However, Point 5 assumes that a particular numerical standard has been chosen and relates to confidence in terms of assessing compliance with that particular standard. There may be different degrees of confidence for different types of action. For example, high confidence (e.g., >95%) might be demanded if controversial or expensive action is being imposed on a discharger. Alternatively, if one is awarding "prizes" for the guaranteed achievement of a standard, 95% confidence that the standard was met might be required. In contrast, if the purpose is only to select sites for further studies or to prioritize monitoring or regulatory activities, a lower degree of confidence might be used (e.g., >50%). The principle in operation here is that monitoring is much cheaper than cleanup, and it is unreasonable to impose severe action on the unwilling solely because others choose not to take enough samples to demonstrate failure with sufficient confidence.

3.3.2　Compliance Issues

An important principle that underpins the use of the 5-point ideal standard is that checking compliance with numerical standards is a statistical process. This requires that a standard is defined in terms that allow the use of statistical methods, typically annual averages and annual percentiles, although these may be applied to spatial as well as temporal variations (e.g., in the assessment of contaminated land or soil quality).

Absolute limits cannot be used in statistical assessments of compliance based on sampling; such standards need to be translated into percentiles to be used. The reasons for this are that standards are usually set in a precautionary manner, and it is nearly always the case that occasional exceedances of a limit value are acceptable. With such an absolute limit, the risk of reporting failure is strongly influenced by sampling frequency: the more sampling, the higher the likelihood that at least 1 sample will fail. This is an important consideration if compliance assessments are to be used to compare regions or nations in a way that has serious implications for "poor" performers.

To illustrate this, consider a watercourse that exceeds a threshold 1% of the time. Such a watercourse will always be reported as a failure of the absolute limit if assessed using continuous, error-free monitoring. Table 3.2 shows that if assessed by sampling, this failure will escape detection with a probability that depends strongly on the number of samples.[2] With 4 samples, there is 4% probability that at least 1 exceeds the absolute limit. This rises to 41% for 52 samples. In this situation, the illusion of improved performance can be manufactured by taking fewer samples. In the meantime, the true quality might have deteriorated.

Absolute limits monitored solely by sampling are not true absolute limits. This is because of the mathematical implication that exceedance is permitted at times (or places) when the samples are not taken; that is, exceedance is tolerated for a proportion of the time (or places). Such absolute limits are, in truth, percentiles.

An additional problem lies in setting up the action to secure compliance. An absolute limit must never be exceeded, but what is meant by "never"? Once in a million years? The answer to this question defines the absolute limit as a percentile; it expresses the proportion of time for which failure is acceptable. This is also the outcome when we select a sampling frequency. There are 31 million seconds in a year, and we might sample 12 of them. This leaves a lot of time when we might fail the absolute limit and not notice it, and this accepted proportion of time defines the percentile: Twelve samples a year treats an absolute limit as something like an annual 95th percentile.

The statistical process sets up a sampling protocol that is constructed to provide an unbiased estimate of the summary statistic that defines the standard, say the annual mean or annual 95th percentile. The statistical methods provide an estimate of the standard error about this estimate of the summary statistic. Usually, the sampling regime

---

TABLE 3.2

Effect of sampling rate on reported compliance

Number of samples	Probability of reporting failure (%)
4	4
12	11
52	41

involves random sampling over the period, or area, covered by the standard (e.g., a year). This is usually structured to impose representation of seasonal effects and not times of maximum risk and assumes that the main variability is temporal. Parallel approaches are required if spatial variability is important. In all cases, monitoring must be set up to provide an unbiased estimate of the summary statistic that defines the standard.

It is expected that Points 1 to 3 concerning the ideal standards will be generated and guided by the scientific considerations of the standard-setting process (see Chapters 4 and 5), while social and economic considerations will have a greater effect on Points 4 and 5 (see Chapter 2).

Although the decision associated with the application of a standard is a decision with a pass or fail value, the application of a standard implies recognition of uncertainty underlying its derivation and in the assessment of compliance. The exceedance of a speed limit by a few miles per hour does not usually lead to prosecution. This may be because the standard is precautionary or because the measuring device is not 100% accurate. The police might choose to act only when the limit is exceeded by, say, 10%, a policy that reflects issues of confidence and precision and the consequences of the violation in terms of the risk of accidents. At greater exceedances of the speed limit, more serious consequences for the offender are likely; therefore, the absolute limit value of the speed limit is not necessarily implemented as an absolute limit that if exceeded will result in equivalent consequences.

For environmental standards, a strict analogy with speed limits can be misleading because the speed limit is an absolute limit, and the driver can secure compliance by driving everywhere just below the limit. In environmental decisions, the situation is more like the case of a new kind of vehicle that needs no driver and whose speed varies randomly and uncontrollably. How should the car be designed so that the speed limit is never exceeded and so that we could check for compliance by observing it, say, once a month?

This might be done by setting the standard at an average speed that is so high that the speed limit is never failed even for 1 second in a year, which implies knowledge of the performance of the car — the relation between average speeds and the highest speeds. In this combination of circumstances, the average standard works well and provides the benefit of huge savings in monitoring because the estimate of the mean is statistically efficient in terms of the cost of securing precise estimates.

An environmental example of a 5-point ideal standard is the following: "In the long run, the 95th percentile value of the concentration of *Escherichia coli* at the chosen monitoring point for a coastal bathing water must be less than 200 organisms mL^{-1} in 19 samples out of 20. Action is initiated when failure is demonstrated with 95% confidence in any summer." In this case, "action" is likely to be a study to check the need to review permits or reduce other sources of contamination. Such action is then designed to secure compliance, in the long run, in 19 summers in 20.

The definition of the summary statistic (in this case, the 95th percentile) leads to the option of — or opportunity for — a statistical assessment of compliance and so provides the facility to estimate the standard error in this estimate and the statistical confidence of failure. This is crucial because compliance is assessed by taking samples, a process associated with high errors and the consequent risk of misdirected action.

It is customary to require 95% confidence of failure for important decisions, or those with expensive consequences, because it is seldom acceptable to act on low levels of confidence and impose large costs on others when this is due solely to a refusal by the regulator to monitor with sufficient frequency. Increased confidence in detecting particular degrees of failure is provided by increased monitoring.

All standards that are used to take decisions will end up with definitions of the 5 points. If these are not specified by the provider of the standard, such as a scientist or risk assessor, the regulator or decision maker will still need to do it. This gives the regulator a free hand in deciding on the stringency of the standard, which is probably undesirable and may cut across the science underpinning the standard. If the 5 points are not defined explicitly, because they relate to the limit value for the substance in question they will vary at each location and for each decision, effectively applying a different (and arbitrary) standard at each location until the unfairness of this is noticed by those having to pay.

It is useful if the considerations of the ideal standard are reflected in the legal documents that support policy on standards or in the guidance that supports the way such documents are used. Standards used technically as percentiles should be declared or interpreted as such in permits and regulations. This removes the risk that a standard used as an annual average or an annual 95th percentile is interpreted by lawyers as an absolute limit.

3.3.3 APPLICATION OF IDEAL STANDARDS TO OTHER MEDIA

For air, soils, sediments, groundwater, lakes, and marine waters, the first point for the ideal standard can be generalized to

- A limit that refers to spatial considerations, such as a concentration of 10 mg kg^{-1}. This might be a spatial average (or an estimate of this) over a specified depth, area, or volume. Alternatively, it might be a proportion of such; for example, the concentration exceeded by 10% of the area. It might be intended that this is estimated by sampling or monitoring, calculated by modeling, estimated by a combination of both, or calculated by using models supported by monitoring that is designed to verify the model. Whatever option is chosen, it needs to be described somewhere. It is important to note that each environmental compartment has its own opportunities for simplifying or complicating the way in which a limit is expressed. For example, in a river it might be that a single spot sample is a reasonable estimate of an average at the time of sampling, with respect to variation with depth or cross section, which simplifies the situation. In contrast, for soil contamination several samples are needed from each different soil type in any given averaging area (cf. CLR 7, DEFRA, and EA 2002) for a definition of averaging area.
- A summary statistic for these spatial concentrations with respect to time or space. For example, how often this limit may be exceeded (say, 5% of the time) or to establish there is no spatial trend. This might be a trivial step in cases of no temporal variation or true spatial heterogeneity.

The air quality management community recognizes the important year-to-year variations in atmospheric concentrations produced by annual variations in weather conditions. However, in broad terms the national and European objectives do not emphasize design risk and statistical confidence as important requirements of an ideal standard. A failure of the annual mean would be regarded as a failure, and similarly for the 24-hour mean standard, although it could be argued that the permitted number of exceedances already take design risk and statistical confidence into account. Derogations are also permitted if high concentration levels are caused by natural phenomena such as Saharan dust, which is a significant factor in Mediterranean countries. The EU Air Quality Directive objectives are subject to uncertainties relating to the effect and the way air quality is measured. Measurement of PM2.5 particles is thought to be more representative than PM10 of the associated health effect. The tapered element oscillating microbalance (TEOM) instrument is widely used in the United Kingdom, which provides an hourly average measurement, but the European Union approved method is a gravimetric weighing scale, and a simple conversion factor is not always appropriate (commonly a conversion factor of 1.3 is proposed).

In 2006, the UK government released its new air quality strategy for consultation (DEFRA 2006). This is the 4th document in a series of strategy reviews, starting with the National Air Quality Strategy in 1997, which set up a series of objectives for 8 pollutants (sulfur dioxide, nitrogen dioxide, particles as PM10, benzene, 1,3-butadiene, ozone, carbon monoxide, and lead). The aim of the strategy was to ensure that air quality met or aimed to achieve these objectives. The revised strategy of January 2000 contained revised objectives and the strategy addendum of February 2003 introduced tighter objectives and different objectives in different parts of the United Kingdom. The objectives are a mixture of long-term standards expressed as annual means and short-term standards, such as one of the standards for particles (PM10), which is a 24-hour mean standard. The objective under the current EU directive is that 50 μg m^{-3} should not be exceeded more than 35 times per year by 1 January 2005, and the objective is that 50 μg m^{-3} should not be exceeded more than 7 times per year by 31 December 2010.

A feature of the present UK system for air quality is that there are a number of tighter indicative standards that do not have the legal force of EU directives. These may be looked on as Type A standards chosen to bring about improvement. However, failure to meet these UK indicative standards has caused public concern because it appears that the government (and Europe) are relaxing standards if these indicative values are not implemented. This illustrates a problem if the role of Type A and Type B standards is not clearly explained. The draft Air Quality Strategy proposes new Type A and Type B standards, with the tighter Type A standards designed to bring about improvement but over a longer timescale and in a way more closely related to the health effect. Extensive analysis and evaluation have gone into national strategy and European air quality evaluations but perhaps without enough attention paid to the way such standards would be implemented.

3.4   OTHER IMPLEMENTATION ISSUES

3.4.1   GEOGRAPHICAL SCOPE OF STANDARDS

The geographical scope of a standard may need to be given consideration, especially when spatial variations exist in factors affecting the toxicity limit value of a standard. This may be through a desire to protect particular species, which may be rare or of particular ecological, economic, or recreational importance to a local area or region, or through the need to take into consideration factors that have a significant effect on the exposure of a substance, such as the bioavailability of some trace metals.

3.4.2   BACKGROUND LEVELS

A distinction sometimes needs to be made between the natural background level of a substance, which arises purely as a result of natural processes, and the ambient background level, which is the concentration measurable in the environment at a "pristine" site (see Section 5.10). In practice, a pristine site is often considered to be one that does not receive any direct inputs from local sources, although it needs to be accepted that for many substances there may be an appreciable input from diffuse atmospheric sources.

The forensic aspect of differentiating natural and anthropogenic contaminants is significant, and the difference in terms of risk posed is sometimes irrelevant. However, it is also fair to say that (metal) contamination tends to be less bioavailable if it is naturally occurring.

Some standards systems (e.g., the proposed UK Environment Agency soil screening values for ecological risk and the proposed EC Soil Framework Directive) have adopted a Dutch approach called the added risk approach, in which it is assumed that the natural background level of a substance cannot have a detrimental effect on an ecological community. With such an approach, the limit value of the standard is added to the natural background level, resulting in the standard that is applied for regulatory or decision-making purposes. This is a policy-based decision, and there are many difficulties in defining the natural background concentration on an appropriate scale. When the background concentration is to be based on ambient background level data, the definition of the background will include an input from diffuse sources of pollution, making it important to determine the "background" ranges of values clearly. Other legislation has specifically chosen not to differentiate between natural and humanmade contamination (e.g., Part 2A of the UK Environmental Protection Act 1990).

Although the scientific basis for taking background concentrations into account may be questioned (e.g., the origin of a substance cannot be distinguished by the affected organism or by analysis), there may be instances when approaches such as added risk are considered for use as pragmatic risk management tools.

3.4.3   FORCING TECHNOLOGICAL INNOVATION

There may be occasions when a standard is set at a concentration below current analytical limits of detection (LODs) or limits of quantification (LOQs). This could be because high uncertainty leads to the application of large assessment factors to toxicity data to derive a standard or because analytical techniques for a particular environmental matrix have higher LODs/LOQs than those available for the medium in which the standard was derived (e.g., sewage effluent versus laboratory water). An inability to measure concentrations of a chemical at the standard does not necessarily render the standard totally useless. For example, a water quality standard set in a receiving watercourse may be below the LOD/LOQ, but measurement of concentrations from an effluent may be above these limits. Appropriate modeling may allow good estimation of whether the standard in the watercourse has been exceeded.

However, it is likely that in most cases standards set at levels below detection and quantification limits should be regarded as considerably less useful than those set above such limits. They should therefore fall toward the screening or tentative end rather than the mandatory end of the standards spectrum. It may be appropriate, in some cases, to use a detectable concentration as a cause for concern or a trigger for remedial action, although in such cases it is important that the standards and how they are implemented are reviewed as the analytical LODs are gradually improved to the point at which the detectable levels may not necessarily be unacceptable. Also, the required accuracy and precision of the method of analysis to be used should be specified to make sure that all those who are potentially affected are being regulated on an equivalent basis.

The aim should usually be to develop appropriate analytical techniques as soon as is practicable for any substance that cannot currently be determined at the concentration represented by the standard.

3.4.4   VERIFICATION AND REVIEW OF STANDARDS

An assessment of whether a particular standard has met its original protection objectives should always be an integral part of the standard-setting process. This should be conducted along with a review of the standard should there be information that can be used to update it and increase overall confidence in its application. A stringent standard may be more likely to meet its protection goals, but it is also important to consider the social and economic aspects surrounding its implementation as it is also important that standards do not place unnecessary burdens beyond what may be required to achieve objectives.

Derivation of a standard for a toxic chemical is often based on extrapolation from a few species to many, from laboratory to field conditions, from high- to low-level effects, and from short- to long-term exposure. A newly derived standard may therefore be a rather uncertain tool for environmental management. Often, this uncertainty is dealt with at the derivation stage using large safety, assessment, or uncertainty factors so that the resulting standard is highly precautionary.

In light of this, it seems wise to build regular review into the implementation strategy for a standard so that users can determine whether the standard is achieving the desired policy outcome without unnecessary (and perhaps costly) stringency. For example, a new standard for protecting aquatic life might be reviewed 3 years after its introduction to determine whether exceedance of the standard is associated with any adverse effects on biologically relevant receptors. By biologically relevant, we mean that if, for example, a standard has been derived on the basis of laboratory toxicity tests on salmonid fish reproduction, then field data verification should be based on the same taxa and endpoints. The usual principles of environmental epidemiology should apply to such correlational analyses to maximize inferential strength (Crane and Babut 2007; Crane et al. 2007). Such a standard may be set to improve the quality and quantity of fish stocks, and it may therefore be appropriate to assess whether these overarching objectives have been achieved.

In undertaking such verification of the usefulness and effectiveness of standards, it is necessary to take account of the potential for many interferences and confounding factors that affect real ecosystems. The ecosystems may be affected by climate, habitat, and contaminants other than those of interest to the study, thus making in situ or field assessment difficult. It is important, wherever possible, to avoid simply concluding that, regardless of whether the standard is passed or failed, environmental or ecological quality does not appear to be affected.

Another form of verification of a standard that does not depend on possibly noncausal correlations between exposure to a chemical and biological effect is the use of controlled experiments such as aquatic mesocosms (see Section 4.7) and experimental field plots. Although such systems do suffer from certain drawbacks (e.g., Crane 1997), they may provide a valuable line of evidence if designed appropriately.

Different lines of evidence from controlled laboratory, mesocosm, and field experiments and from uncontrolled field-based correlations can be used during periodic reviews in a weight-of-evidence approach to determine whether a standard is fit for use or should be relaxed or tightened.

3.4.5 IMPLEMENTATION ANALYSIS REPORT

We argue in this chapter that an implementation analysis should inform the entire life cycle of a standard, from the original specification, through derivation, to its implementation, use, review, and eventual revision or withdrawal. It will also be helpful to anticipate future uses such as the use of compliance assessments to compare regions and nations (where these are not in place already).

A separate implementation analysis report is therefore unnecessary because the narrative that describes the standard should include all of the elements necessary for its implementation (i.e., its magnitude, duration, and frequency [to inform the summary statistic by which the standard is defined], the design risk for remediation, and significance level for determining failure for each type of decision). Indeed, separating implementation analysis from other aspects of standard setting is likely to lead to more standards that are not used; to standards that are used inconsistently from place to place; to remediation measures that are laxer or stricter than actually needed to meet the standard or achieve the protection goals; and to biased reports of the relative performance of regions and nations.

Early consideration of implementation options during the specification stage will ensure that resources are not wasted on deriving standards that cannot be used and will help to focus effort on the best solutions.

NOTES

1. Available from: http://www.ec.gc.ca/ceqg-rcqe/english/default.cfm.
2. These calculations are based on the probability of no failures in a set of N samples. This is 1 minus P to the power N, where P is the probability of a compliant sample. Thus, for 12 samples in Table 3.2 it is 1 minus 0.99 raised to the power 12.

REFERENCES

Crane M. 1997. Research needs for predictive multispecies tests in aquatic toxicology. Hydrobiologia 346:149–155.

Crane M, Babut M. 2007. Environmental quality standards for Water Framework Directive priority substances: challenges and opportunities. Integrated Environ Assess Manage 3:290–296.

Crane M, Kwok KWH, Wells C, Whitehouse P. 2007. Use of field data to support European Water Framework Directive quality standards for dissolved metals. Environ Sci Technol. 41:5014–5021.

[DEFRA] Department for Environment, Food and Rural Affairs. 2006 April. The air quality strategy for England, Scotland, Wales and Northern Ireland. A consultation document on options for further improvements in air quality, volume 1. London.

[DEFRA and EA] Department for Environment, Food and Rural Affairs and Environment Agency, UK. 2002. CLR 7: assessment of risks to human health from land contamination: an overview of the development of the soil guideline values and related research. Bristol (UK).

[EC] European Commission. 2006. Proposal for a directive of the European Parliament and of the Council on Environmental Quality Standards in the Field of Water Policy and Amending Directive 2000/60/EC. Brussels: COM(2006) 397 final, 17 July 2006.

[ICRCL] Interdepartmental Committee on the Redevelopment of Contaminated Land. 1987. Guidance on the assessment and redevelopment of contaminated land. 2nd ed. London: Department of the Environment London. ICRCL Guidance Note 59/83 [now withdrawn].

Nathanail CP. 2005. Generic and site specific assessment criteria in human health risk assessment of contaminated soil. J Soil Use Manage 21:500–507.

[RCEP] Royal Commission on Environmental Pollution. 1998. Setting environmental standards. 21st report. London: HMSO. Available from: http://www.rcep.org.uk/reports/21-standards/document. Accessed 23 June 2009.

[USEPA] US Environmental Protection Agency. 2000. Methodology for deriving ambient water quality criteria for the protection of human health. EPA-822-B-00-004, WEPA, Washington, DC, USA.

4 Water and Sediment EQS Derivation and Application

Peter Matthiessen, Marc Babut,
Graeme Batley, Mark Douglas, John Fawell,
Udo Hommen, Thomas H. Hutchinson,
Martien Janssen, Dawn Maycock,
Mary Reiley, Uwe Schneider,
and Lennart Weltje

4.1 INTRODUCTION

This chapter deals with the derivation of aquatic environmental quality standards (EQSs), including standards for the protection of water dwellers, predators of water dwellers, and human water users. However, the main focus is on standards for the protection of organisms that live in water or aquatic sediment and are able to absorb contaminants directly via their gills, skin, and/or cell surfaces. In other words, the chapter primarily covers the derivation of standards for the protection of aquatic ecosystems.

The objective of aquatic EQSs is to assist environmental managers in either maintaining or improving the quality of a water body by identifying the chemical contamination that might lead to or might have caused deterioration, quantifying its severity, and locating the source of damage. It is best not to use EQSs in isolation; ideally, they should be used as lines of evidence in an environmental assessment. There will usually need to be a range of EQSs available to assist the attainment of a range of protection goals for different water bodies or sections of water bodies. Thus, aquatic EQSs may well include values to protect freshwater and/or marine pelagic ecosystems, freshwater and/or marine benthic ecosystems, and/or the food chain dependent on aquatic organisms. There may also be a need to consider amenities such as bathing or drinking water quality, although in some regions drinking water quality is considered through a different regulatory process. There is a need to be protective of raw water as a potential source of drinking water abstraction (even though it will usually then be subject to treatment), and in certain countries this is achieved through the setting of aquatic EQSs (e.g., US Environmental Protection Agency [USEPA] 2000; Lepper 2005).

If several EQSs apply to a given site, the aim should generally be to attain the lowest concentration (Lepper 2005) or most protective use (USEPA 1994), although this may be overridden by policy considerations. EQSs used for protecting aquatic ecosystems should generally be focused on the protection of their structure (i.e., their populations and communities). This will usually tend to protect ecosystem function. In Europe, for example, this is reflected in the use of species diversity as an important tool for assessing the ecological quality of surface waters (Allan et al. 2006). However, from a microbial perspective, it makes more sense to focus directly on ecosystem function, although this is usually only done in experimental situations (e.g., activated sludge and sediment–water biodegradation tests; OECD 1984).

Many countries generate a single EQS value for a given substance. In some cases, this may be modified at a local level to take account of site-specific factors (e.g., in Canada). However, in recognition of the uncertainty that inevitably surrounds the derivation and application of EQS values (e.g., due to the need to extrapolate from limited and variable data sets and the need to recognize the degrees of variability in natural populations), it is recommended that each EQS should be framed as a range of at least 2 values representing higher and lower levels of protection. An example of this is the Australian approach (ANZECC/ARMCANZ 2000), in which separate EQS values are derived to cover pristine, moderately disturbed, and highly disturbed ecosystems. In this case, the EQS values are based on species protection levels of 99%, 95%, and 90%, respectively, but other procedures are also possible (e.g., the threshold effects level/probable effects level approach used for sediments in some parts of the United States, Canada, and elsewhere (CCME 1999a; Batley et al. 2005). Similarly, in France, for freshwater quality assessment, there are 5 surface water classes representing pristine to highly degraded systems (Oudin and Maupas 2003). At least theoretically, the classes are based on structure and abundance of species; the EQSs are then based on no observed effect concentration (NOEC) and/or EC50 values modified by a range of assessment factors (AFs) (Babut et al. 2003). A similar approach is being developed in the United States (USEPA 2005c).

The approaches described generate a matrix of EQS values for a given substance, 1 axis of which covers different protection goals and 1 of which represents values with different levels of protection (see Table 4.1). This is currently only applied to ecosystem protection guidelines for waters, although in some jurisdictions, 2 levels are applied to sediments. It should be noted that while habitats of currently low quality might only be expected initially to comply with EQS values giving an "acceptable" degree of protection, they could (and arguably should, where feasible) be targeted for remediation to the point at which "good" or "excellent" EQS values are applied.

Although most circumstances require the use of chronic or long-term EQSs (e.g., annual average concentrations, AAs) to achieve optimal environmental protection, there is also a need for acute or short-term values (e.g., maximum acceptable concentrations, MACs) for use in evaluating the potential consequences of transient events such as spills and pesticide runoff occurrences. There is also a need to be able to assess drinking water contamination emergencies (e.g., resulting from diesel spillage) in a pragmatic manner. The way in which the short-term values are deployed will depend on operational matters such as sampling frequency. Short- and long-term EQS derivation is discussed in more detail in Section 4.4.

TABLE 4.1

Example of a matrix of potential water EQS values for a given substance

	Protection goals				
Level of protection	Water ecosystem	Sediment ecosystem	Food chain	Recreation	Drinking water
High	EQS-1	EQS-4	EQS-7	EQS-10	EQS-13
Medium	EQS-2	EQS-5	EQS-8	EQS-11	EQS-14
Acceptable	EQS-3	EQS-6	EQS-9	EQS-12	EQS-15

Several environmental factors (e.g., temperature, salinity, pH, hardness, dissolved organic matter [DOM]) can alter the bioavailability and hence the toxicity of some water contaminants. For example, dissolved copper concentrations in many industrialized estuaries exceed EQS values but are thought rarely to cause damage because they tend to be associated with high concentrations of DOM, and hence bioavailability is reduced (Sunda and Guillard 1976). If scientific knowledge permits, EQS values should therefore be accompanied by guidance on how to interpret them in the light of specified water quality conditions. Examples of this include the approaches to ammonia, which take account of the modifying factors of temperature and pH (e.g., USEPA 1999; ANZECC/ARMCANZ 2000; CCME 2000); the proposed US approach to copper (USEPA 2003c); and the proposed EU approach to cadmium, which takes account of hardness (Lepper 2005). It should be noted that measuring the bioavailable fraction of a contaminant under operational conditions may be difficult (or impossible) in many instances and expensive (predicting it would be cheaper, but may be subject to significant uncertainties), but assessing compliance in this manner may avoid costly remediation options that would be required on the basis of measured total contaminant concentrations.

When employed in monitoring discharges, long-term EQSs are often enforced at the edge of a mixing zone around the discharge (e.g., USEPA 1994). However, depending on the degree of environmental protection required, one can also consider applying EQSs immediately downstream of the discharge. In such cases, it might be more appropriate to use the short-term EQS (USEPA 1994) if expected dilution in the receiving water is satisfactory. On the other hand, when assessing environmental quality in areas more remote from discharges, one would generally use long-term EQS values.

While the EQS approach is rightly recognized as providing a valuable environmental assessment tool, it is important to realize that an EQS exceedance (particularly a small one of no more than a factor of 2 for a short-term EQS or substantially less than 2 for a long-term EQS) does not necessarily imply that environmental damage is occurring or that immediate regulatory action is required. Similarly, compliance with an EQS does not necessarily ensure protection of all local ecosystems. Although EQS failures alone may be useful for ranking the quality of different locations and can lead to immediate regulatory action in some situations, in many cases they should trigger an investigation that could involve such techniques as ecotoxicity testing of water or sediment samples; in situ toxicity tests; toxicity identification and evaluation (TIE);

ecological studies of plant, fish, or invertebrate communities; or biomarker measurements for specific classes of microcontaminants (e.g., fish vitellogenin for estrogens; Vermeirssen et al. 2005). These should be used to assess the weight of evidence that the environment is being damaged and that a suspect substance is indeed the cause of degradation. This process of investigation should proceed in a tiered manner to minimize costs and maximize explanatory power (e.g., Canadian Water Quality Guidelines, CCME 2007; Simpson et al. 2005; Government of Canada 2006).

4.2 SPECIFICATION AND RECORDING OF EQS DERIVATION PROCEDURES

There needs to be a balance between the use of prescriptive EQS derivation procedures to encourage consistency and the need for evidence-based judgment in complex scientific areas (e.g., when using nonstandard or incomplete data). EQS derivation procedures have been specified in a variety of ways, ranging from those driven largely by evidence-based judgment (Zabel and Cole 1999) to prescriptive methods with less scope for interpretation (e.g., CCME 1991; Lepper 2005). Although the use of transparent, evidence-based judgment can be very effective (e.g., the Zabel and Cole approach was used for EQS derivation until recently in the United Kingdom), the need for consistency (especially when deriving standards to be used internationally) has rightly driven the majority of regulators down a more prescriptive route. Prescriptive methods, as well as evidence-based judgments, should nevertheless be accompanied by transparency of decision making and a detailed audit trail (i.e., full records kept of reasons for each decision).

In some jurisdictions (e.g., United States), EQS derivation methods used by central governments may be very prescriptive but permit local adjustments to account for particular species of higher or lower concern (Stephan et al.1985; USEPA 1994). These flexible arrangements will generally require evidence-based decision making and documentation at a local level. Overprescription, however, carries the risk of straitjacketed decision making and may lead to the derivation of unrealistic EQS values in some circumstances. Furthermore, the prescriptive use of large AFs can lead to unnecessarily low EQSs without the "reality check" provided by experts familiar with the behavior and effects of a substance in the field. It is therefore now recognized in most jurisdictions that increased prescription should nevertheless embrace the use of evidence-based judgment in defined circumstances. For example, although there is little need for expert input when assessing standard acute toxicity data sets, that need increases considerably for higher-tier data such as mesocosm responses and impacts on endocrine systems and when considering chemical speciation and impacts of other physicochemical modifying factors. Evidence-based decision making may also be needed to screen complex data for quality.

In summary, it is recommended that while EQS derivation procedures should be carefully and explicitly prescribed, evidence-based judgment should also be used for screening and assessing higher-tier studies, unusual endpoints, complex test designs, outlying data, and so on (e.g., see the USEPA assessment of atrazine, USEPA 2003a, and the Canadian use of nonstandard endpoints relating to salmon and fluoride,

CCME 2002, available from: http://www.ec.gc.ca/ceqg-rcqe/English/Pdf/GAAG_ Fluoride_e.pdf). This may need to be followed up by the verification of unusual and potentially critical data through repeated testing. It is also essential that evidence-based decision making when deriving EQSs is fully recorded, transparent, and as consistent as possible from case to case. "The draft EQS should then be subject to scientific expert peer review followed by public consultation."

4.3 SELECTION AND EVALUATION OF DATA FOR DERIVING WATER AND SEDIMENT EQSs

4.3.1 SELECTION AND PRIORITIZATION OF SUBSTANCES FOR EQS SETTING

The selection and prioritization of substances for which EQSs are required is generally based on both scientific and political criteria. Scientific criteria include the intensity of use of a substance and its occurrence in the environment (i.e., the likelihood of aquatic exposure) as well as information about its (eco)toxicological properties. Thus, a strong driver might be if a substance belongs to the group of persistent, bioaccumulative, and toxic (PBT) chemicals. Examples can be found in the COMMPS procedure (Fraunhofer IUCT 1999) as well as in the Australian, Canadian, and US approaches, all of which are summarized in Table 4.2.

4.3.2 TYPES OF DATA

4.3.2.1 Species Selection

Data requirements differ between countries and regions. While the EU approach based on the technical guidance document on chemical risk assessment (TGD) (EC 2003) can start with very limited data to which high AFs are applied, Canada, the United States, and Australia require a more elaborate minimum data set (or all acceptable data available, whichever is the greater) and do not apply high AFs if more data are available, as the species sensitivity distribution (SSD) approach accounts for the decreased uncertainty as n increases. As discussed in Section 4.4.1.2, the use of SSDs is now preferred over the previously used standard test species approach if the data are adequate. We consider that the EU acute base set (algae, *Daphnia*, fish) is insufficient for definitive EQS setting. More data (especially chronic data and data on more species) are considered necessary, but the EU base set may be appropriate for deriving screening or "tentative" values.

A question remains about the minimum number of species that are necessary to adequately represent aquatic taxa for purposes of EQS derivation. There is no consensus among the authors about this minimum data requirement.

4.3.2.2 Test Endpoints

In general, standard tests for effects on survival, development, reproduction, and growth are used for EQS derivation. Other endpoints should be used if they are considered relevant at the population level (e.g., behavior and avoidance). It is currently not considered appropriate to use biomarker data (e.g., molecular and biochemical endpoints) for EQS derivation unless their relevance at the population level has been clearly demonstrated.

TABLE 4.2

Examples of approaches in different jurisdictions to substance prioritization for EQS derivation

	Australia	Canada	European Union	United States	WHO (drinking water)
Science drivers — national and local priorities	Based on high toxicity, persistence, and bioaccumulation potential; coupled with high usage and detection in environmental compartments; substances whose EQSs appear overly conservative (exceeded at control sites) are targeted for reevaluation	Cross-media occurrence; inherently toxic; persistent; bioaccumulation potential; presence on the Canadian market; likelihood of environmental release; detection in the environment at concentrations that may cause harm; recent advances in the environmental toxicology of the substance	Monitoring data driven (with use of models if monitoring data not available); high toxicity; PBT criteria used to obtain further ranking as a priority hazard substance	Based on frequency of occurrence in the environment; availability of toxicity data demonstrating effects; opportunity for exposure; criteria as needed to protect the chemical, physical, and biological integrity of US waters	Evidence for reasonably widespread actual or potential occurrence in drinking water at levels that may be of health significance, combined with evidence of actual or potential toxicity
Policy drivers — international, national, or local	General concern for internationally recognized priority chemicals; specialized national issues based on identifying the significance of international priority substances (e.g., dioxins); local issues relating to either toxic incidents or new chemical registration applications	New information on substances with existing EQSs; use of the EQS despite being outdated and therefore affecting policies inappropriately; current environmental monitoring of the substance; interim status of the current EQS; availability of EQSs for the substance in other media; substance identified by other prioritization schemes within Canada or	• International agreements (e.g., Rhine Commission [before adoption of WFD]); extended monitoring followed by screening risk assessment • Legislation such as *Water Framework Directive* (*COMPPS Procedure*) (Fraunhofer IUCT 1999) • Existing national lists of substances to be	Chemicals or other pollutants determined to be of concern because of likely effects on threatened and endangered species or on commercially or recreationally important species or of ecosystem function significance; Existing Priority Pollutant List, chemicals or pollutants identified as significant contributors to	Significant international concern

	ANZECC/ARMCANZ (2000)	CCME (1991) Environment Canada	European Commission (1999)	Clean Water Act Sec. 304(a)	WHO (2003, 2006)
Stakeholder drivers	Stakeholders invited to provide listings of substances of concern; drivers might be compounds detected in industrial discharges; awareness of international concerns (e.g., endocrine disruptors, pharmaceuticals); inputs from industry, NGOs, public	internationally; an EQS would assist in the mandated deliverables of Environment Canada (its operational requirements and strategic projects). An EQS would assist in the mandated deliverables of other departments; need to meet binational agreements on the substance; need to support concerns of territorial, provincial, or municipal jurisdictions; need to support operational requirements of the private sector; need to support operational requirements of other countries. Could an EQS be derived in partnership with other stakeholders within Canada or internationally? Are there external funding contributions?	monitored under various pieces of legislation (e.g., EU Directive 76/464)	the impairment of waters in the United States per the Clean Water Act Sec. 303d list of impaired waters. States and tribes may identify a need for an EQS: industry and NGOs may identify a need for an EQS or have an interest; Congress may identify needs or interests; the media may also uncover potential problems that an EQS could address	Not applicable
Reference	ANZECC/ARMCANZ (2000)	CCME (1991) Environment Canada (http://www.ec.gc.ca/ceqg-rcqe/English/Ceqg/Water/default.cfm) accessed 23 June 2009)	European Commission (1999)	Clean Water Act Sec. 304(a)	WHO (2003, 2006)

Chronic toxicity data are preferred for deriving an annual average EQS (AA-EQS). Acute data are used to calculate a maximum acceptable concentration EQS (MAC-EQS) and can be used to derive the AA-EQS if insufficient chronic data are available, but an AA-EQS should not be derived exclusively on the basis of acute data. Guides to efficient decision making about the testing requirements for derivation of short- and long-term EQSs based on modes of action and other considerations were provided by Verhaar et al. (1992), de Wolf et al. (2005), and Hutchinson et al. (2006).

A range of other data is often required when deriving aquatic EQS values, particularly when deriving AA-EQSs. Examples of the data required by the EU Water Framework Directive (WFD) are shown in Table 4.3. Most of these are also required under other jurisdictions.

4.3.2.3 Algal Tests

Results of standard algal tests are sometimes handled differently from ecotoxicological tests with animals due to the short generation times of algae. For example, under the Uniform Principles of EU Directive 91/414 (risk assessment for plant protection products), the EC50 of a standard algal test is used with an AF of 10, while an AF of 100 is used for acute EC50 or LC50 data from invertebrates or fish. In addition, the EC50 of an algal or *Lemna* test is used in the same way as a NOEC from chronic studies with *Daphnia* or fish. The reason for this approach is that an algal test can be interpreted as a chronic study if the generation time of the algae is taken into account. Due to their fast population growth, the recovery potential of algae is expected to be high. In addition, it can be argued that when considering the protection of algae, their function is more important than their community structure (i.e., they are similar to microbial communities). In contrast to the EU Uniform Principles for pesticides, which treat algal tests as chronic, algae are handled in the same way as other test organisms under the EQS derivation framework of the EU TGD and WFD, that is, an AF of 100 is applied to the lowest EC50 of algae, *Daphnia,* and fish to calculate an MAC-EQS.

Regarding algal growth inhibition testing (which often gives an ErC50 [EC50 estimated from specific growth rate] and EbC50 [EC50 estimated from biomass growth]), several authors have argued that the biomass endpoint should not be used (reviewed in Section 6.3.2.1 of the TGD; European Commission [EC] 2003). The reason is that direct use of the biomass concentration without logarithmic transformation cannot be applied to an analysis of results from a system in exponential growth. If only the EbC50 is reported but primary data are available, a reanalysis of the data should therefore be carried out to determine the ErC50. If only the EbC50 is reported and no primary test data are available, consideration should be given to performing a new algal study to obtain a valid ErC50 and NOEC or ErC10.

The algal growth inhibition test is not only a multigeneration test but also provides a measure of sublethal effect — reduction in population growth. In Europe, it is therefore considered a true chronic test (i.e., EC10 or NOEC based on algal growth rate) (EC 2003). As a result of these considerations, we recommend that the results of algal growth tests should be treated as chronic data for the purposes of EQS derivation. Canada is currently reviewing the use of algal tests, and there is a possibility that extremely short tests (1 to 4 hours) will be classified as acute data for the purposes of EQS derivation.

4.3.2.4 Sediment-Dwelling Organisms

If no tests with sediment-dwelling organisms are available, data on organisms living in water may be used for some substances to derive tentative sediment values by means of equilibrium partitioning (EP) but should not be used for definitive sediment EQSs. In principle, it is also possible to calculate a tentative water EQS from a sediment EQS, although this is rarely done because water EQSs are more plentiful than those for sediments. Although EP appears useful for estimating the safe sediment concentrations of very hydrophobic substances, the present state of science and limited data imply that caution should generally be exercised before extrapolating sediment EQS values from water EQS values. At present, experimental sediment toxicity studies are considered necessary as a basis for deriving sediment EQSs (see Batley et al. 2005).

4.3.3 DATA VALIDITY

The data must be checked for validity (e.g., Klimisch et al. 1997; Durda and Preziosi 2000). Data used should

- Be reliable and relevant (e.g., generated according to test guidelines)
- Have been assessed using proper statistical analysis methods
- Avoid unrealistically high test concentrations that may create artifacts such as accumulation of undissolved particulates on fish gills
- Be based on experiments in which test concentrations were measured, with measured concentrations used to define endpoints if they differ from nominals by more than ±20%
- Be fully documented (e.g., conducted to Good Laboratory Practice (GLP))
- Have clear dose–response relationships (except for tests in which the substance's water solubility does not permit the testing of higher concentrations and consequently the test endpoint is equal to or higher than the highest test concentration or in cases of a limit test for which only 1 relatively high concentration was tested and no effects were found)

If data have already been generated for risk assessments, it may be possible to use these without further quality assessment (as is done in the European Union), but any new and relevant information should be considered scientifically. Usually, the lowest acute and long-term toxicity data for a species should be used (if the endpoint is relevant at the population level and is derived from a validated study). If valid data for the same endpoints and the same species exist, the geometric mean should be taken. This value is then used together with other species values to calculate the EQSs. Use of a mean value in this situation minimizes "cumulative conservatism"; that is, always using the most conservative data point, especially when additional valid data are available, will lead to too much conservatism, giving rise to an EQS that may be unrealistically low.

4.3.4 DATA REQUIREMENTS OF DIFFERENT EQS ASSESSMENT METHODS

Full details of these assessment methods are provided in Section 4.4. The data required by various jurisdictions are listed in Tables 4.3 to 4.5.

TABLE 4.3

Other data that may be required for deriving EQSs under the EU Water Framework Directive (also required by most other regulatory authorities)

Data type	Comments
Multispecies, microcosm, and mesocosm studies	Can also be used if they fulfill quality criteria and are interpreted in the context of EQS setting
NOEC or NOEL from long-term avian and mammalian feeding studies	Required to assess secondary poisoning of wildlife. These studies are preferred to acute data.
Mammalian toxicity data (oral toxicity, repeated dose toxicity, carcinogenicity, mutagenicity, and reproduction studies)	Required for assessing possible food chain effects on human health
QSAR outputs	These can be used as supporting data but should not be used exclusively for EQS derivation.
Kow, experimental bioconcentration factor (BCF) or bioaccumulation factors (BAFs), field BAFs, or field biomagnification factors (BMFs)	Required to assess bioaccumulation potential and possible effects on the food chain
Data on persistence in waters, sediments, and tissues	These are not used directly for EQS derivation but are useful supplementary information for use in risk management and the like.
Partition coefficients (e.g., Kow, Ksusp–water, Ksed–water, Koc)	May be required for transformation calculations (e.g., equilibrium partitioning)
Other supplementary data (e.g., water solubility, vapor pressure, and molecular weight)	Used for plausibility checks etc.
Data on background levels of metals	These are required when calculating metal EQS values in the European Union and Canada but not in the United States, where site-specific metals bioavailability is used to calculate the value and account for unique site characteristics (see Section 4.10).

A hierarchy of data requirements, which is an efficient way to approach the derivation of EQSs, can be considered for application in the following methods:

- Use of a small group of standard data sets with fixed AFs (standard test species approach; see Brock et al. 2006 and Section 4.4.1.1)
- Species sensitivity distribution approach (Section 4.4.1.2)
- Model ecosystem approach (see Sections 4.4.1.3 and 4.7)

While more data will generally help derive a more robust aquatic EQS, the reality is that for many substances data are few. As indicated, the standard test species approach using AFs can be applied to very small data sets (e.g., toxicity data on 1 alga, 1 crustacean, and 1 fish), but it is not recommended for deriving definitive EQSs.

TABLE 4.4

Summary of ecotoxicological data requirements used in different jurisdictions for aquatic (surface water) EQS derivation

	Australia	Canada	European Union	United States
Short-term EQS (acute lethal data)	No short-term EQSs derived	Freshwater: • 3 invertebrates • 3 fish Saltwater: • 2 invertebrates • 3 fish	L(E)C50 for at least the base set (algae,[a] Daphnia, fish) or others. For SSD, L(E)C50 for at least 10 species, including fish, 2nd family of Chordata, crustacean, insect, a family of a phylum other than Arthropoda or Chordata (e.g., Rotifera, Annelida, Mollusca, etc.), a family in any order of insects or any phylum not already represented, alga, higher plants (data for marine and freshwater species may be combined if reasonable), or others. NOEC based on initial concentration of model ecosystem study	• 1 salmonid • 1 other fish • 1 chordate • 1 planktonic crustacean • 1 benthic crustacean • 1 insect larva • 1 rotifer, annelid, or mollusc • 1 other insect larva or mollusc
Long-term EQS (chronic data, e.g., growth and reproduction)	Using the SSD approach, chronic NOEC data are required for at least 5 different species from at least 4 taxonomic groups. Using the application factor approach, at least 5 data points representing the	Freshwater • 3 fish • 3 invertebrates • 1 plant Saltwater • 3 fish • 2 invertebrates • 1 macrophyte/algae	1) Long-term EQS for pelagic communities of inland surface waters: Acute base set (algae, Daphnia, fish) requiring AF = 1000, 1 to 3 long-term NOECs from the base set requiring AF 100 - 10, at least 10 NOECs (for species, see short-term SSD), or field data or model ecosystem data 2) Long-term EQS for pelagic communities of	Same species as for acute, or 3 species and an acute–chronic ratio The same numbers of organisms are required for marine assessments, but water-type appropriate species should be used.

(continued)

TABLE 4.4

Summary of ecotoxicological data requirements used in different jurisdictions for aquatic (surface water) EQS derivation (continued)

	Australia	Canada	European Union	United States
	basic trophic levels, aquatic plants, crustaceans, and fish (1 fish, 1 alga or aquatic plant, and 2 invertebrates). Separate marine and freshwater EQSs are derived with priority given to using the SSD approach.		transitional or coastal waters: Usually the same data are used as for inland waters, but more stringent AFs if marine species are not adequately represented, such as NOECs from 3 freshwater or saltwater species representing 3 trophic levels plus 2 NOECs from additional marine taxonomic groups (e.g., echinoderms, molluscs) would require an AF of 10.	Both freshwater and saltwater numbers are published if available. SSDs are always used if sufficient data are available.
Number of aquatic EQS derivations completed to date	85 freshwater; 26 marine	200	33 under WFD	46
Reference	ANZECC/ARMCANZ (2000)	CCME (1991)	Lepper (2005)	National recommended WQC table (http://www.epa.gov/waterscience) Stephan et al. (1985)

[a] Note that short-term algal toxicity studies, strictly speaking, should be regarded as chronic tests because they encompass several generation times.

TABLE 4.5

Summary of ecotoxicological data requirements used in different jurisdictions for sediment EQS setting

	Australia	Canada	European Union	United States
Short-term sediment EQS	No short-term EQSs	No short-term EQSs	No data are required as transient peak levels in sediment are not expected.	Water criteria value or OC for nonionic organics or use AVS-SEM for metal mixtures Protocol under development for tissue-based criteria (e.g., selenium)
Long-term sediment EQS	No fixed requirements yet. EQSs currently based on US effects database; new approaches are being developed (e.g., Simpson and Batley 2007) No separate freshwater guidelines, but application mainly to marine systems	Freshwater: At least 4 studies on 2 or more sediment-resident invertebrate species, including at least • 1 benthic crustacean • 1 benthic arthropod other than crustacean At least 2 of these studies must be partial or full life-cycle tests with ecologically relevant endpoints. Marine: At least 4 studies on 2 or more	If tests with sediment organisms are available, same requirements as for pelagic communities; otherwise, equilibrium partitioning method, such as up to 3 NOECs or EC10 values for species representing different living and feeding conditions, to which are applied AFs of 100, 50, or 10; same approach for marine sediments but more stringent AFs if marine organisms are not adequately represented (see Table 4.3)	Same protocols using appropriate species

(continued)

TABLE 4.5

Summary of ecotoxicological data requirements used in different jurisdictions for sediment EQS setting (continued)

	Australia	Canada	European Union	United States
		sediment-resident invertebrate species, including at least • 1 benthic amphipod At least 2 of these studies must be partial or full life-cycle tests with ecologically relevant endpoints. For derivation of a definitive EQS, must have at least 20 entries in the BEDS database		Methodologies for metal mixtures, dieldrin, endrin, and PAH mixtures have been published with example calculations and values.
Number of sediment EQS derivations completed to date	34 interim guidelines	—	Approximately 20, mainly based on equilibrium partitioning	
Reference	ANZECC/ARMCANZ (2000)	CCME (1999a)	Lepper (2005) and environmental quality standards (EQSs) Substance data sheets for the EU priority substances (available at http://forum.europa.eu.int/ Public/irc/env/wfd/library?l= /framework_directive/i-priority_ substances/supporting_ substances/ background/ substance_ sheets&vm= detailed&sb= Title)	*Procedures for the Derivation of Equilibrium Partitioning Sediment Benchmarks (ESBs) for the Protection of Benthic Organisms.* Consult http:// www.epa.gov/nheerl/ publications/ for these procedures for PAH mixtures, dieldrin, endrin, and heavy metals.

For the SSD approach, data requirements are more demanding and variable (see Table 4.6). Although acute data can be used to produce SSDs, the resulting output cannot be used directly for AA-EQS derivation, for which chronic data are generally required. For example, in the Netherlands SSDs are calculated if at least 4 chronic NOECs are available, while according to the EU TGD and the WFD, chronic data for at least 10 species covering different groups of taxa should be used, which is unrealistic for many substances. For plant protection products, the workshop on Higher-tier Aquatic Risk Assessment for Pesticides (HARAP) recommended basing SSDs on data for at least 8 species from the most sensitive group, except for fish, for which 5 species are considered to be sufficient for ethical reasons (Campbell et al. 1999) and because fish are a much more homogeneous group than algae or aquatic invertebrates. Goodness-of-fit tests and visual inspection are required to check quality of the fit. Note that the Australian approach uses a generalized statistical fit rather than the forced log-logistic fit employed elsewhere (ANZECC/ARMCANZ 2000) (see also Section 4.4.1.2).

For substances with specific modes of action, the SSD should be based on data for the most sensitive group (e.g., crustaceans). If there are no indications that salinity affects toxicity (see Section 4.6), SSDs can be created using combined freshwater

TABLE 4.6
Comparison of assessment factors in current use

Regulatory authority	Data set	Assessment factor	Notes
Canada[a]	Chronic LOEC	10	3 species of fish, 2 invertebrates, and algae (or freshwater vascular plant)
Australia	Lowest of 5 or more species chronic NOECs	10	If necessary, NOECs are estimated from other data as follows: MATC[b]/2 LOEC/2.5 E(L)C50/5
	Lowest of 5 or more species acute LC(EC)50	100 or 10 × ACR	Applied to lowest
European Union (freshwater)	Acute LC(EC)50	1000	
	Chronic NOEC	10 to 100	
European Union (marine)	Acute LC(EC)50	1000 to 10 000	
	Chronic NOEC	10 to 1000	
OECD	Chronic NOEC	10 to 100	

[a] CCME 2007.
[b] MATC, maximum acceptable toxicant concentration = (NOEC + LOEC)/2.

and marine data sets (although attempts to do this in the United States have not been successful because of stakeholder preferences for water-specific values).

The mesocosm assessment approach is described in Section 4.7.

4.3.5 Use of Toxic Body Burdens for Assessing Sediment Toxicity

Although still in the research phase, there is some international interest in using the contaminant body burdens of sediment dwellers to help derive sediment EQS values, that is, to define a "safe" sediment in terms of bioaccumulated body burdens (lethal internal exposure concentrations) instead of, or in addition to, deriving a safe concentration for the sediment itself.

This approach gets around the issue of bioavailability of sedimentary contaminants. There are published examples for metals (Simpson and Batley 2007) and organochlorines. However, it should be noted that the approach can be difficult for essential metals whose concentrations in tissues are typically regulated over a range of ambient concentrations. Furthermore, there may be technical difficulties involved in measuring the body burdens of sediment dwellers.

4.4 DERIVATION OF EQSs

4.4.1 Available Methods

Predicted no-effect concentrations (PNECs) are traditionally used in risk assessments and form the basis of most EQS values. There are 3 basic approaches to the derivation of PNECs and resultant EQS values. The traditional method uses standard toxicity data and applies an AF to the most sensitive endpoint to derive a protective concentration. The SSD approach utilizes all available toxicity data to derive a value that is protective of a given percentage (e.g., 95%) of the species and is increasingly being used by many countries, often with a small AF placed on, for example, a predicted HC5 (hazardous concentration to 5% of species, i.e., the 5th percentile of the SSD) based on chronic data. Finally, model ecosystems such as lentic mesocosms can be used to derive safe values, again usually with a small AF.

4.4.1.1 Standard Test Species Approach

The standard test species approach uses either acute LC(EC)50 data or chronic data, which usually take the form of the NOEC but can also be expressed as an ECx, indicating a low- or no-effect threshold.

From both statistical and biological points of view, the use of NOECs has some shortcomings compared to ECx values (OECD 2006a):

- Since the NOEC (or no-observed-effect level, NOEL) does not estimate a model parameter, a confidence interval cannot be assessed.
- The value of the NOEC is limited to being 1 of the tested concentrations (i.e., if different values are chosen for the tested concentrations, the value of the NOEC will change).

- If the power is low (due to high variability in the measured response and/or small sample size), biologically important differences between the control and treatment groups may not be identified as significantly different. If the power is high, it may be found that biologically unimportant differences are found to be statistically significantly different (OECD 2006a).
- In addition, the size of the effect at the Lowest Observed Effect Concentration (LOEC) might vary considerably from test to test.
- As a consequence, some authors (e.g., Chapman et al. 1996) recommend the use of ECx values instead of NOECs in the evaluation of ecotoxicological tests. Often, the EC10 is considered as a reasonable alternative to the NOEC (e.g., EU WFD, Lepper 2005).

It should, however, be noted that NOECs do have some advantages, including their use with studies having only a few dose levels (e.g., some mesocosm trials); with endpoints (e.g., behavior) not amenable to the ECx approach; with effects that vary in severity as well as occurrence; or with endpoints with no underlying dose–response relationship.

There are various approaches that have been adopted for the use of AFs with the available toxicity data. They differ from authority to authority in the species number and type required and in the factors that are applicable to the given toxicity data (acute versus chronic). The various factors are summarized in Table 4.6. These factors reflect uncertainty in intra- and interlaboratory variation, intra- and interspecies variation, extrapolation between short- and long-term toxicity, extrapolation of results from laboratory to field, the possibility of indirect effects such as interspecific reactions (e.g., loss of predators, affecting prey), and the fact that environmental contaminants are often present as complex mixtures.

Application of AFs should also be considered for mesocosm data (see Section 4.7.2.1).

As discussed in Section 4.8, and providing that "some" chronic data are available, use of acute-to-chronic ratios (ACRs) can be helpful if chronic data are limited to allow use of the large acute toxicity database for PNEC or EQS derivation.

$$ACR = Acute\ LC(EC)50/Chronic\ NOEC$$

It is emphasized again that one needs several chronic data sets from the taxonomic group in question to obtain a reliable ACR, in which case it can be more appropriate to use the chronic data directly for EQS derivation if sufficient data are available (in the United States, there would have to be 8 data sets).

These discussions at present apply mainly to toxicity data derived from water-dwelling organisms, although in principle they could also apply to sediment toxicity data. Experience is limited with respect to the application of AFs to sediment data, the main problems being the restricted number of sediment test species available and the need to consider multiple exposure pathways.

4.4.1.2   Species Sensitivity Distributions

There has been a large increase in recent years in the application of SSDs in ecotoxicology, as evidenced by the recent SETAC (Society of Environmental Toxicology and Chemistry) book by Posthuma et al. (2002). This approach is being or will shortly be applied in the European Union, Australia, United States, and Canada, with a general movement (where possible) toward the use of ECx in preference to the historical use of AFs and NOECs in deterministic PNEC calculations. The toxicity endpoint used will depend on the objective. For deriving a MAC-EQS, LC(EC)50 data are appropriate, while chronic NOECs (or preferably chronic ECx values, where available) are applicable for AA-EQSs.

The minimum data requirements vary with different countries as discussed (e.g., Australia uses 5 species from 4 taxonomic groups, with at least 1 fish, 2 invertebrates, and 1 alga or plant; ANZECC/ARMCANZ 2000), while the European Union uses at least 10 species (preferably more than 15) covering at least 8 taxonomic groups when assessing existing substances (EC 2003; Lepper 2005). Note that the USEPA differs slightly in using the 4 lowest genus mean acute values (GMAVs), from a minimum of 8, to obtain the freshwater acute value (FAV), then divides by the ACR to obtain the freshwater chronic value (FCV) (Stephan et al. 1985). It should also be noted that SSDs that include unrelated taxonomic groups are not appropriate for substances with specific modes of action.

The more data that are available for the SSD, the greater the reliability of the derived EQS will be. The main advantage of the SSD approach is that it provides a confidence interval for a given predicted safe value (e.g., an HC5 or concentration predicted to be in exceedance of a particular endpoint, such as NOEC or LC50, for 5% of species in the relevant group). For plant protection products, Brock et al. (2006) suggested using the 95th percentile of the HC5 for EQS setting, while the median HC5 can be used for setting the more relaxed regulatory acceptable concentration (RAC). These different concentrations are related to different protection principles, the "ecological threshold principle," which can be considered the principle behind standard setting for large water bodies (e.g., under the WFD) or sites with important nature conservation functions, and the "community recovery principle," which might be appropriate for multifunctional aquatic ecosystems adjacent to discharges and the like. This gives rise to a range of potential EQSs for a given substance (see Table 4.1).

Data fit is also critical, and statistical tests for goodness of fit should be applied. The log-normal distribution is used in Europe (Aldenberg and Jaworska 2000) and the log-logistic in the United States, while Australia uses a Burr-type distribution (ANZECC/ARMCANZ 2000; Shao 2000), and Canada is also considering the use of a range of statistical distributions to get the best fit. Not unexpectedly, the type of statistical model chosen will have some effect on the resulting HC5 (see Table 4.7).

The SSD approach has the added benefit that the percentage of species protected can be altered to derive EQS values applicable for different levels of protection, such as 99%, 95%, and 90% (i.e., HC1, HC5, and HC10, respectively), for pristine, slightly or moderately disturbed, and highly disturbed habitats, as used in Australia

TABLE 4.7

Effect of statistical model on derived HC5 values obtained in Australia for metals[a]

Metal	No. of data points	HC5 µg L⁻¹	
		Log-logistic	Burr type III
Zinc	17	8.0	5.0
Cadmium	21	0.10	0.19
Copper	21	0.9	1.4
Nickel	6	6.2	11.1

[a] While precise goodness of fit cannot be provided for these data, all fits were acceptable. Note also that no judgment is implied about which distribution is "better" in a particular case.

(see Table 4.4). These percentage levels of protection are associated in Australia with 50% confidence limits, consistent with the Dutch approach. Use of 95% confidence limits is common elsewhere but is more limited in scope due to the higher data requirements that it imposes, and it can lead to overly conservative values that are often below background concentrations (see Section 4.9).

There have been numerous studies examining the selection of data for an SSD. Forbes and Calow (2002) made the point that only a fraction of the species going into the SSD determines the effects threshold. With all species being weighted equally, the loss of any species is of equal importance to the system, while keystone or other important species are assumed to be randomly distributed in the SSD. For example, the ecologically realistic distribution of species by trophic level was 64% primary producers, 26% herbivores (invertebrates), and 10% carnivores (fish), compared to the mean ratio from SSDs for different chemicals of 27.5, 34.7, and 37.8%, respectively. Such variations were shown to alter the SSDs by as much as 10% (Duboudin et al. 2004). A sensitivity analysis performed on available data for chromium (VI) in marine waters (Table 4.8) shows how additional data points, or selective removal of data, have an impact on the derived 5th percentile (HC5). The effects are relatively small but can be higher for the 1st percentile data (HC1). Our view is that, provided the data set includes numbers of sensitive and insensitive species equal to or above the minimum data set, it is considered to be adequate.

Data selection criteria have been discussed (Section 4.3). Maltby et al. (2005), in a discussion of the use of tropical or temperate species and lentic or lotic species, deemed it appropriate to include all in the 1 SSD assessment. However, if a substance has a specific mode of action (like a pesticide), it is generally agreed that the SSD should only include taxa from the most sensitive group (e.g., just arthropods for an insecticide). It is not appropriate to include unrelated insensitive taxa in such a distribution (e.g., algae in an SSD for an insecticide) since it will result in a bi- or trimodal distribution and thus a poor fit.

TABLE 4.8

Effect of data selection on SSD-derived EQS values (HC1 and HC5) for chromium (VI) in marine waters[a]

Data set used	HC5 µg L⁻¹	HC1 µg L⁻¹
Full data set (43 points)	6.8	0.30
Full data set but lowest data not geometric mean for each species	3.1	0.13
Full set less a duplicated low point	10.0	0.56
Full data set less 2 low points	14.6	1.1
The previous set with 3 high points added	15.9	1.1
Australian EQS value	4.4	0.14

[a] Data from ANZECC/ARMCANZ 2000 database.

In the EU risk assessment TGD (EC 2003), an AF between 1 and 5 is suggested for HC5s derived from SSDs (based on chronic NOECs) to obtain a PNEC. The size of the AF should be based on, for example, the quality of the data set, the endpoints covered, the diversity of the species tested, the goodness of fit, and sampling uncertainty (e.g., the confidence interval around the HC5) and comparison made with results of model ecosystem or field studies. In the framework of pesticide registration in Europe, no recommendation on AFs is made, but it is stated that the standard AF can be reduced by up to 1 order of magnitude if additional species have been tested (Santé des Consommateurs [SANCO] 2002). In the United States, the Aquatic Life Guidelines provide that the calculated value can be decreased to be protective of a commercially or recreationally important or other named species as the means to provide protection for species that are found to be more sensitive than the value provided by extrapolation from the 4 lowest GMAVs (Stephan et al. 1985).

Our consensus is that SSDs are preferable to the standard test species approach when deriving PNECs if there are sufficient laboratory data. However, users must be aware of potential difficulties, including the need for care in the choice of suitable data, in the use of appropriate statistical models, and in the use of AFs. Furthermore, well-conducted model ecosystem studies may provide more ecological realism and less uncertainty than an SSD.

4.4.1.3 Predictions from Model Ecosystem (Microcosm and Mesocosm) Data

Predictions of PNECs or EQSs based on mesocosm or microcosm chronic NOEC data, if available, are probably of greatest reliability, and such data can be used as a benchmark for comparison with other EQS derivation methods (see Section 4.7).

After analyzing data sets for 16 pesticides, as well as assessing evidence from the open literature, Brock et al. (2006, p. 25) showed that

> The median acute HC5 value is lower than the effect class 2 concentration observed in microcosm and mesocosm experiments treated once with a pesticide. The corresponding lower-limit HC5 value was, with high certainty, lower than reported effect class 1

and effect class 2 concentrations, and even when the pesticide was applied repeatedly the median acute HC5 is lower than the effect class 2 concentrations.

In other words, SSD predictions tend toward overprotection for acute effects in some cases.[1]

4.4.2 SELECTION OF THE MOST APPROPRIATE PNEC OR EQS DERIVATION METHOD

One can rank the reliability of EQSs according to the method of derivation, as shown in Table 4.9 for long-term EQSs. Ideally, proposed EQSs should be accompanied by an explicit assessment of their reliability.

Note that it has been shown that confidence in the proposed EQS increases with the number of taxa used (as the number of sensitive species included will tend to be greater). There is also a need to consider embryonic or larval stages of fish and invertebrates (McKim 1977) because they are often more sensitive than the corresponding juvenile or adult forms.

4.4.3 CORRECTIONS FOR BIOAVAILABILITY

Bioavailability should be considered when assessing compliance of a water body with an EQS, rather than modifying the derivation of the EQS as discussed, unless the EQS is calculated or designed to be calculated on a site-specific basis. Increasingly in the United States, EQSs are being presented as equations or models to be used on a site-specific basis rather than as a national number to be applied broadly; thus, bioavailability is addressed within the derivation of the value rather than at the time of compliance comparison.

When constructing an SSD, it is important to use data for the same range of test conditions (e.g., pH, salinity). For example, PNECs derived for metals should ideally

---

TABLE 4.9
Reliability classification of long-term EQSs[a]

Reliability	Derivation method
High	Model ecosystem data with small assessment factor
	SSDs based on chronic NOECs (or ECx) with minimum data set or greater plus small or no assessment factor
	Medium assessment factor applied to lowest chronic NOEC from data set of 5 or more species
Medium	SSDs based on acute data with large assessment factor
	Lowest acute LC(EC)50 data for 5 or more species with large assessment factor
Low	Small acute data set with very large assessment factor; small chronic data set with large assessment factor

[a] A similar ranking of reliability can be developed for short-term EQSs.

be calculated from tests using synthetic waters or pristine waters with low dissolved organic carbon (DOC). Measured concentrations in the tests should then be corrected for bioavailability in accordance with the known water quality. Such corrections for the effects of DOC on the acute toxicity of metals can be made by speciation measurement or modeling (e.g., using the biotic ligand model; DiToro et al. 2001) or based on toxicity testing.

As a word of caution, it should be noted that the finding in Australia that hardness does not affect the response to copper of the most sensitive biota (algae, cladocerans) has meant that the validity of a hardness correction to the EQS that was derived mainly from fish data has been brought into question (Markich et al. 2005). Similar findings have been made in Europe regarding the hardness correction for zinc toxicity.

4.5 SHORT- (MAC) AND LONG-TERM (AA) EQSs AND IMPLICATIONS OF EXCEEDANCE

Long-term AA-EQSs are intended to protect against "long-term continuous exposure," and MAC-EQSs are intended to protect against "short-term episodic events." It should be noted that the use of acute toxicity data for deriving AA-EQSs is considered permissible by some jurisdictions, for example, when suitable chronic toxicity data are not available (e.g., CCME, 2007), although the use of chronic data for this purpose is always preferable.

Confounding the issue is the frequent confusion of the terms "acute" and "chronic" with "short term" and "long term." In general, the terms "acute" and "chronic" should be used to describe the types of effects observed or likely to occur in relation to the life cycle of the species tested, whereas the terms "short term" and "long term" should be used to delineate the duration of exposure.

The AA-EQSs in principle apply to all substances (to guard against the possibility of continuous or semicontinuous exposure, e.g., from manufacturing plant discharges), while it is arguable that MAC-EQSs provide an "additional" standard solely for transient substances. However, persistent substances still need to be regulated for acute episodic events, and transient substances may be released continuously over longer periods (e.g., manufacturing discharges). Consequently, as it is envisaged that both scenarios can occur for all substances, AA-EQSs and MAC-EQSs are potentially required for all.

The distinction between AA-EQSs and MAC-EQSs has little meaning without specification of the monitoring regime. For example, it is clearly inappropriate to compare single measurements of environmental concentration with AA-EQSs derived from longer-term exposures (Solomon et al. 1996). In the United States, 4-day time-weighted averages (TWAs) are used for short-term assessments of effluent or river quality and 30-day TWAs for long-term assessments, although these methods tend to be data intensive. They may, however, be more relevant for ecological assessments than an annual average comprised of 12 monthly samples evenly spaced throughout the year unless the substance is continually released into the environment.

The difficulty comes with the number of samples collected. If only 1 sample is analyzed in the month, then that measured value is, by default, both the peak value and the TWA, which can lead to an overprecautionary assessment. Since the results of the routine monthly sample are unlikely to be available until well after the event, this prevents the opportunity to take additional samples to determine a more accurate 30-day TWA, although it has to be recognized that more intensive routine monitoring schemes may be unrealistically expensive. Exceedance of an EQS value (particularly a single exceedance of a MAC-EQS) does not automatically demonstrate ecosystem damage but indicates a potential for damage that might depend on the degree of exceedance. For these reasons, exceedance of an EQS should not necessarily lead to regulatory action but should act as a trigger for further investigation. While EQS values are internationally recognized as a valuable environmental assessment tool, the uncertainty concerning their ecological realism must always be borne in mind.

We support the concept of standards expressed as percentiles, such as "x mg L^{-1} must not be exceeded for $y\%$ of time during the year" rather than as absolute thresholds, and this theme is developed further in Chapter 3 (see also USEPA 1991, 2005a).

A clear recommendation is that the actual EQS applying to a particular substance at a particular site should be appropriate for the specific exposure scenario under consideration. The sampling regime should therefore ideally be tailored to the type of substance and its discharge pattern. As sampling strategy and application of EQSs currently vary enormously between countries, our recommendations about sampling and the application of EQSs are recognized to have significant operational and regulatory implications on the international scale.

4.6 MARINE AND FRESHWATER EQS DERIVATION

4.6.1 NEED FOR SEPARATE MARINE AND FRESHWATER STANDARDS

There are 2 primary issues to consider regarding the need for separate marine and freshwater standards:

1) The possibility that there may be differential exposure effects caused by salinity
2) The possibility that some marine organisms show a different sensitivity to contaminants from freshwater organisms

The higher ionic strength of marine waters (chloride and carbonate concentrations), as well as higher pH, controls the speciation or dissociation of inorganic chemicals or polar organic compounds. Solubility of polar organics and metals will be salinity or pH dependent. Moreover, adsorption to particles or solubility in colloids may also be influenced by salinity (Means 1995). For species living in these marine environments, salinity also influences their physiology, so responses to toxic stress are likely to be different (Table 4.10). For example, pyrethroid insecticides are more toxic to some marine organisms (e.g., arthropods) than freshwater organisms, most probably due to their ion-channel-disrupting activity (Solomon et al. 2001).

TABLE 4.10

Examples of organic and inorganic substances for which toxicity is different in freshwater and marine water organisms

Substance	Differential sensitivity	Reference
Ammonia	Markedly greater sensitivity claimed for freshwater species; however, the 2 data sets on which this is based are not comparable: freshwater fish (34%) and crustaceans (26%) versus saltwater crustaceans (72%), no annelids or platyhelminths	Wheeler et al. 2002
Pyrethroid insecticides	Marine arthropods and fish more sensitive than freshwater equivalents	Solomon et al. 2001
Organometallics	The bioavailability of tributyltin at neutral or alkaline pH is elevated, leading to an increased toxicity to molluscs.	Rüdel 2003; USEPA 2003b
Metals	Lower FW HC5 values for copper, lead, nickel, zinc, and potassium dichromate to a range of species; no differences in sensitivity for cadmium and only a slight differential for mercury.	Wheeler et al. 2002

There are few data allowing comparisons of sensitivities of species in marine and freshwaters (Leung et al. 2001). The concept of sensitivity might also be questioned as it often refers to a comparison of various species' responses in standardized bioassays, which may not represent actual exposure routes; to some extent, a more sensitive species could be considered simply to be more exposed or exposed for longer. There are some indications that algae, crustaceans, and fish are of more or less similar sensitivity in fresh- and marine water, although there are some contradictory examples (see, e.g., a review for atrazine in Solomon et al. 1996 and Figure 4.1), but mollusc and echinoderm embryos may be more sensitive to a range of organics than many freshwater species. Although taxa living in marine waters or estuaries are deemed more diverse than freshwater species (Russell and Yonge 1928; Tait 1978), the supporting evidence is not abundant. Moss (1988) stated that 56 phyla are present in marine waters compared to 41 in freshwaters, but no phyla are confined to freshwaters only, while 15 phyla are found only in marine waters. The MarSens project (Peters et al. 2006) concluded that marine organisms can be significantly more sensitive to narcotics when the analysis is conducted on the taxonomic level of phyla and, while calling for more data, tentatively recommended applying an additional AF of 10 to marine toxicity information. Peters et al. also concluded (without providing a reference) that marine biodiversity is higher than freshwater, with 78% of aquatic species being marine. Therefore, it is assumed in some jurisdictions (e.g., European Union) that marine sensitivity distributions will encompass a wider range than in freshwater systems, thus justifying higher AFs (EC 2003). It might, however, be better to base such decisions on differences in the shape and position of marine and freshwater SSDs and not on biodiversity per se. On the other hand, marine

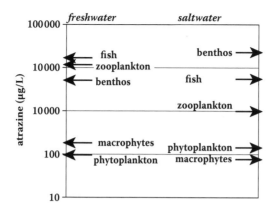

FIGURE 4.1 Geometric means of acute atrazine toxicity data in freshwater and saltwater organisms. (Reprinted with permission from Solomon et al. 1996 © Society of Environmental Toxicology and Chemistry.)

waters tend to be more homogeneous than freshwaters in terms of physicochemical conditions such as DOM, carbonate, pH, and so on.

Wheeler et al. (2002) established acute freshwater and saltwater SSDs for 21 substances, including ammonia, metals, several pesticides, and narcotic substances. Using HC5 calculations and curve slope, they found freshwater species were either more sensitive (ammonia, copper, nickel, or zinc) or less sensitive (chlordane, endosulfan, pentachlorophenol) than saltwater species. In some cases, the distributions were very similar; however, the taxonomic compositions of the freshwater and saltwater data sets were not always comparable. Maltby et al. (2005) analyzed SSDs for 16 insecticides and inter alia compared SSDs based on saltwater and freshwater species. They concluded (page 379) that the "taxonomic composition of the species assemblage used to construct the SSD does have a significant influence on the assessment of hazard, but the habitat and geographical distribution of the species do not." Differences in freshwater and saltwater SSDs were primarily driven by taxonomy (e.g., both crustaceans and insects are present in freshwater, but only crustaceans are found in seawater). Correcting for the disparity in taxonomy removed habitat differences.

Overall, the lack of data hampers a sound and definitive comparison, but current scientific opinion is that there is no "systematic" bias in sensitivity between freshwater and marine species, provided similar tests and endpoints are involved. Nevertheless, a direct read-across from freshwater to marine data should not be made automatically. The argument for higher marine AFs is therefore not entirely clear-cut, although their use may be appropriate in some cases. We agree with Lepper (2005) that an additional AF should not automatically be applied to marine data, but only if the available data do not adequately represent marine life in the ecosystem to be protected. Furthermore, if there is no indication of differential sensitivity to a particular substance between freshwater and marine organisms, it may be appropriate to combine both data sets in a single SSD, although any resulting EQS should be regarded as tentative.

4.6.2 MARINE TOXICITY DATA REQUIREMENTS

As a whole, marine data requirements are similar to those for freshwater in terms of number of taxa and phyla to be tested. Tests are available for marine or estuarine species and for species able to support a large range of salinities. Widely used standard tests published by the International Organization for Standardization (ISO), ASTM International (formerly ASTM, American Society for Testing and Materials), OSPAR Commission, USEPA, and others cover a wide range of groups, including algae, annelids, crustaceans, echinoderms, molluscs, and fish.

4.6.3 SUBSTITUTION OF FRESHWATER FOR MARINE DATA (AND VICE VERSA)

Given the possible differences in sensitivity of freshwater and marine species, substitution from 1 data set to the other is not generally recommended except if insufficient freshwater or marine data are available for extrapolation to an EQS.

In practice, there will sometimes be situations for which saltwater toxicity data are not available. In these situations, it may be deemed necessary to use freshwater data in lieu of data for marine species. However, several regulatory authorities (e.g., Australia, Canada, and the United States) would not extrapolate from freshwater data to set a definitive marine EQS. Based on acute SSDs, Wheeler et al. (2002) concluded that it is possible to use freshwater toxicity data to extrapolate to saltwater effects by applying an "appropriate assessment factor." However, they did not propose a value for this factor; moreover, looking in detail at their results (see Section 4.6.1), this extrapolation could be overprotective in some cases.

Our consensus is that if a marine (acute or chronic) EQS is needed, then it should be based on experimental marine data and not extrapolated from freshwater studies. If data are substituted from freshwater systems to marine (or vice versa), the proposed standard should only be considered as "tentative" (see Zabel and Cole 1999). A tentative value of this type is likely to be unreliable if used for regulation.

4.7 USE OF MICROCOSM, MESOCOSM, AND FIELD DATA

4.7.1 INTRODUCTION TO MICROCOSM AND MESOCOSM TESTS

There is a wide spectrum of these microcosm and mesocosm higher-tier test systems available, from laboratory microcosms to outdoor mesocosms, enclosures, and artificial streams (ECETOC 1997). Fish are usually not included, and this is a problem if single-species tests suggest that fish are more sensitive than algae or invertebrates (Girling et al. 2000).

Micro- and mesocosm studies are most often used for higher-tier testing of plant protection products (Campbell et al. 1999; Giddings et al. 2002; Brock et al. 2006; OECD 2006b), employing a single application of test substance rather than maintained test concentrations. Most test systems simulate lentic water (ditches, ponds) because these are considered to represent worst-case exposure situations for periodic entries. These water bodies also receive nutrients from agricultural areas and thus might be eutrophic. However, several such studies have also been conducted with industrial chemicals and more or less constant exposure. Artificial streams are rarely used for pesticides, but some stream studies have been done for industrial chemicals simulating chronic exposure (Belanger et al. 1995; Belanger 1997).

Studies for pesticide risk assessment focus on the situation in a water body near the field edge, with the peak of exposure soon after application due to drift, runoff, or drainage. In most cases, endpoints are related to the initial concentration of the test item (which can encompass multiple applications). In contrast to this, the focus of EQS derivation is mostly on protection against effects of long-term exposure.

Endpoints available to date are most often presented as NOECs and rarely ECx values. Compared to laboratory tests, the number of replicates is usually lower and the variability between replicates higher (due to complexity of the systems and study durations of several weeks or months). Thus, the statistical power might be low, especially for rarer taxa. Due to these limitations, assessment of effects should not be restricted to statistical NOECs alone but also based on inspection of plots of population dynamics. Multivariate statistics are generally used to test for effects on community structure.

For pesticide risk assessment, recovery is taken into account to derive an NOEAEC (no-observed ecologically adverse effect concentration). A maximum of 8 weeks is often taken as an acceptable recovery period, but the life cycle of the affected species should also be taken into consideration (SANCO 2002).

4.7.2 Use of Microcosm, Mesocosm, and Field Studies for EQS Setting

4.7.2.1 Use of Existing Studies

Historically, the United Kingdom and some other regulatory authorities have based EQS values on single-species tests and only used mesocosm data as corroborative evidence. Under the WFD, however, mesocosm no-effect data have recently been used directly to set a freshwater EQS for chlorpyrifos in the European Union, and similar action has been taken for atrazine in the United States (USEPA 2003a). It is clear that such data are more relevant to the natural environment than laboratory-based single-species tests.

Mesocosm no-effect data on a rapidly dissipating compound such as a pyrethroid insecticide may not be suitable for a chronic EQS applied to a river. Furthermore, most existing micro- and mesocosm studies are inappropriate for EQS derivation if fish are the most sensitive species because fish have generally been excluded from such tests. There is consequently a need for evidence-based decision making for interpretation of nonstandard mesocosm studies. Microcosm and mesocosm tests can, however, be used directly for EQS derivation if algae, macrophytes, and invertebrates are appropriately represented in the test systems and if they concern substances subject to transient exposure. They are then directly applicable for the derivation of MAC-EQSs. For this purpose, the NOEAEC can be used as it represents the highest initial concentration that causes no ecologically relevant effects.

Microcosms and mesocosms can also directly be used to derive AA-EQSs if the test substance concentration was relatively constant over time (due to stability of the test item or due to experimental manipulation) and the study duration was long enough to detect possible long-term effects (usually at least 8 weeks, but this will depend on the life cycle of sensitive taxa). These test systems are more difficult to use for AA-EQS derivation if substances have short half-lives in the test system but are expected to be continuously released into water bodies in the field. NOEAECs derived in higher-tier studies for pesticides with short half-lives cannot be used directly as AA-EQSs if they are based on initial concentrations and recovery was the endpoint considered. In this case, reevaluation and interpretation of observed effects are required to calculate the AA-EQS (e.g., relate the highest treatment level without ecologically relevant effects to a TWA concentration).

The need and the criteria for determination of AFs applied to the endpoints of model ecosystem studies (e.g., a community NOEC or NOEAEC) are still under debate. For pesticides, the AF is usually well below 10 but often larger than 1 (SANCO 2002). Criteria for the size of the AF are, for example, the quality of the study. Depending on the "protection principle" and the type of exposure, Brock et al. (2006) suggested an AF of 1 to 5 applied to microcosm and mesocosm endpoints with effects classified as "not demonstrated" or "slight" according to SANCO (2002).

There are a few examples for which field data have been useful for EQS derivation (e.g., the tributyltin EQS in the United Kingdom), but such examples are rare, and the use of field data is generally restricted to a supporting or corroborative role. This is because good cause–effect data are not often available from field studies. However, it is important to keep an open mind about the value of field data for supporting future EQS derivation programs, and studies of pollution gradients can be of particular assistance.

4.7.2.2 Use of New Microcosm and Mesocosm Studies

Microcosm and mesocosm studies can be directly designed for the purpose of EQS derivation (e.g., the exposure scenario, communities to be monitored, etc.). Guidance for design and conduct of microcosm and mesocosm studies can be found in the references given for pesticide risk assessment, but OECD has recently published a guideline for a lentic field test that is not focused on pesticides alone (OECD 2006b).

4.8 CALCULATED ESTIMATES OF TOXICITY

For the derivation of EQSs (and similar benchmarks), experimental toxicity data are considered essential. However, for many substances there will be insufficient reliable toxicity data available to meet the prescribed minimum data requirements. In their absence (or to supplement an existing data set), several extrapolative methods may potentially be of assistance. Nevertheless, we recommend extreme caution when extrapolating from calculated values to predicted real toxicity data. Most suggested calculation methods to supplement missing toxicological data are considered unacceptable in EQS derivation.

Available methods are as follows:

- Quantitative structure–activity relationships (QSARs):
  - QSARs are deemed potentially applicable, especially if the QSAR is extensively validated (see critical reviews of available QSARs for ecotoxicology: European Center for Ecotoxicology and Toxicology of Chemicals [ECETOC] 1998; Comber et al. 2003).

- However, they must only be applied within the group of chemicals represented by the QSAR training set.
- They should work well within groups of similar chemicals (e.g., polychlorinated biphenyls [PCBs] and polyaromatic hydrocarbons [PAHs]).
- These should only be used by experienced assessors to "create" data used in the derivation of a tentative value but must not be used for definitive EQS derivation.
- These could be a useful tool as part of the priority-setting process.
- Acute-to-chronic ratios (see also Section 4.4.1.1):
  - The ACRs are an established method to extrapolate to chronic values from acute data.
  - Several related taxa (e.g., 3) are generally required to maintain scientific defensibility of the approach.
  - The ACRs are extensively used in some regulatory authorities, but others have rejected them.
- Toxicity correlations between taxons (e.g., fish and daphnids)
- Are considered to be potentially useful if sufficiently validated but can only be confidently used in the derivation of a "tentative" value

We recommend that if an extrapolation method is used, then it should be validated, documented, and supported by suitable data.

To properly manage the risks to ecosystems, EQSs for all media are required. If EQSs for related media are derived, the consistency between the EQSs across media should in principle be verified (i.e., water–tissue, water–sediment, sediment–tissue, water–soil, water–air, soil–air) to ensure that a value that is protective for 1 medium is also protective for the other media.

Possible calculation methods for "reading across" between media include

- Equilibrium partitioning for calculating the pore water concentration of hydrophobic organic substances in sediments for comparison with the "safe" value in water (Di Toro et al. 1991).
- The acid volatile sulfide (AVS) approach to calculate the dissolved (pore water) fraction of metals in sediments, again for comparison with the water EQS (Di Toro et al. 1990).
- A sediment biotic ligand model (sBLM) (Di Toro et al. 2005), along the lines of the aquatic BLM (Paquin et al. 2002), has recently been developed to include EP to sediment binding phases; notably sediment organic carbon (OC) only is considered. Application of this first-generation sBLM was able to model organism response within an order of magnitude.

These tools may be useful in assessing consistency of EQSs between media, but as they are not validated for all substances, they can only be applied in limited circumstances at present. In principle, a "tentative value" for some chemicals in sediments might be derived from their water EQSs by such extrapolation methods, but they are not yet sufficiently validated for routine use when calculating definitive sediment EQS values. It might also be valid to extrapolate to a tentative water EQS from a sediment EQS using the same methods.

Experimental data on sediment toxicity are generally required at present for setting a reliable sediment EQS value, although some regulatory authorities are already deriving full EQSs using calculation methods (e.g., Australia has used EP to derive a tributyltin sediment guideline based on the water EQS).

While we recommend caution in the use of calculation methods in place of actual toxicity tests, we recognize the longer-term importance, for ethical reasons, of finding validated alternatives to vertebrate animal testing (i.e., with fish and amphibians), such as QSARS. For the present, however, there is no viable alternative to biological testing when deriving definitive EQS values.

4.9   BACKGROUND CONTAMINATION BY NATURALLY OCCURRING SUBSTANCES

In some jurisdictions (e.g., the European Union and Canada), policy decisions have been made in regard to the modification of aquatic EQSs to account for the natural background of the substance (e.g., for metals). The ways in which this can be achieved are discussed next.

4.9.1   METALS

4.9.1.1   Identifying Background Concentrations

Measured background concentrations of various heavy metals in some locations may be close to, or exceed, existing or proposed regulatory standards (G-BASE 2009). The reasons for this are numerous, including the significant geological variability of metals, the metal concentrations used in control media in toxicity testing, and inappropriate methodologies for standard derivation. For metals for which speciation-based approaches exist (i.e., BLMs for copper, zinc, and nickel), there is no need to account for background concentrations. Similarly, if the background concentration of a metal in water is considerably below the EQSs (i.e., the difference between the Predicted Environmental Concentration (PEC) and PNEC is large), accounting for background will make little difference to the outcome. However, for metals for which the EQS is within the range of likely background concentrations, there is a need to consider how the standards can be modified. This modification is based less on scientific and technical merit than regulatory pragmatism.

In Europe, the notion that it may be possible to determine a "natural background concentration" of a metal at most sites may be unrealistic. Many hundreds of years of industrial activity, urbanization, and widespread aerial deposition mean that it is best to consider background concentrations as those determined at sites of relatively low anthropogenic impact. These concentrations will still vary considerably from site to site due to geological influences.

Similarly, in Canada the "natural background" is defined as that fraction of ambient metal considered to be dependent on the biogeochemistry of the site, in contrast to the fraction that reflects historical human activity. However, in some areas of the world, historical human activity (e.g., mining) stretches back millennia, and the metal contamination derived from this may now be considered as "background" because local fauna have had time to adapt. "An important point related to the background issue, which is sometimes overlooked, is that local EQSs for biologically essential metals must not be set at such a low level that they result in metal deficiency problems."

Theoretically, one can consider both physiological adaptation during an organism's life span and genetic adaptation of a population over several generations as factors that might justify the adoption of particular background levels as safe benchmarks for local setting of EQSs (e.g., see published work on adaptation of marine invertebrates to arsenic, tin, and zinc in the Fal estuary, southwest England by Bryan et al. 1987). It is, of course, important that this approach should not be used as a means of approving current polluting activities but applied solely to environments contaminated by past anthropogenic activities or by the natural geological background. However, this does raise the contentious question of an appropriate "cutoff" in time for when, or when not, to account for the increased background concentration produced by an industrial activity.

In another approach, statistical analysis methods have successfully been used to classify marine sediments; a given element in the fine sediment is normalized to aluminum (which represents the geological background). Points deviating above the expected regression line are considered to reflect levels above natural background (OSPAR Commission 1997). This approach has also been used to establish background concentrations in soils, making use of well-established biogeochemical processes (Hamon et al. 2004; Zhao et al. 2007). Further examples include methods used to discriminate between natural and anthropogenic sources of rare earth elements (Bau and Dulski 1996; Weltje 2002) and lead (Östlund and Sternbeck 2001; Caurant et al. 2006) using isotopic ratios.

In the United States, characterization of the background concentration of a substance is part of the USEPA risk assessment and risk management program for "Superfund" sites[2] (USEPA 1989). The primary goal of the program is to protect human health from the current and potential release of the hazardous substances of concern. Since some of these substances may occur naturally, establishing background levels is important because the CERCLA program does not generally clean up concentrations below natural or anthropogenic background levels. The USEPA (2002) suggests a statistical approach for evaluation of data points based on statistical tests of discordancy; however, Qian and Lyons (2006) have recently put forward a statistical method that is based on a Bayesian approach (estimation) that leads to a smaller bias.

There is a clear need to extend the scope of databases of background metal concentrations in surface waters to allow accurate use of metal EQSs at the local level. For example, the UK and US Geological Surveys (e.g., G-BASE) hold highly detailed databases of total (i.e., bioavailable plus nonbioavailable) metals in streams, and it would be useful if these could be correlated with biological measures of stream quality to identify local metal concentrations that can genuinely be considered as background levels. Unfortunately, the monitoring data from G-BASE in the United Kingdom is not spatially consistent and relies predominantly on small streams and not rivers (which is often where the EQSs are applied).

We are not able to recommend any single approach for identification of background concentrations as preferable to the others.

4.9.1.2 Modifying Metal EQSs to Account for Background

It is clear that if measured concentrations of a metal are close to or above the EQS and the speciation-based approach is not available, then special attention needs to be given to identification of the local background level (however it is defined). In all cases, it is the bioavailable metal concentration that is relevant, but this is often not well defined and not readily determined for data-poor metals.

The total risk approach accounts for the total dissolved metal in a water body, implying that no distinction is made between the ambient fraction of a metal in a water body and the added fraction (Lepper 2005). This approach can result in a standard below the natural background. Hence, a suggestion in the European Union to account for background is the added risk approach. It allows interpretation of the outcome of exposure and effects analysis or risk characterization in terms of the different fractions (i.e., the natural background [total metal]) and the anthropogenic fraction.

Using the added risk approach (Crommentuijn et al. 2000), the amended EQS or maximum permissible concentration (MPC) is defined as follows:

Amended EQS = Background concentration + Maximum permissible addition (MPA)

where the MPA is equal to the PNEC obtained from EQS derivation.

A crucial difference between the total and added risk is in treatment of the ecotoxicity data. The added risk methodology requires the production of a $PNEC_{add}$, which is calculated by subtracting the background concentration of metal in the control from the effect/no-effect concentration from that test. These data are then used for the derivation of the $PNEC_{add}$. The ambient background concentration at the site is then added to the $PNEC_{add}$ to produce the standards to which the measured concentration is then compared.

The added risk approach is based on the assumption that the ecosystem is already "adapted," and has developed, due to the background concentration at the site of interest, and the MPA reflects the anthropogenic addition that the system can tolerate before effects are seen. The flowchart in Figure 4.2 sets out the essentials of the added risk procedure.

Note that in Australia, Canada, and some other jurisdictions, the added risk procedure is not followed, and the EQS instead defaults to the background concentration if this is greater than the calculated EQS at a particular site (ANZECC/ARMCANZ 2000; CCME 2007). In nearly all jurisdictions, speciation and bioavailability considerations are taken into account, but for some metals this is not currently possible.

Generally, it is considered that separate approaches are not needed for taking account of background concentrations in water and sediment.

4.9.2 Other Substances

In principle, the added risk approach could be applied (in the European Union and Canada) to other substances, for example, natural steroidal estrogens from wildlife,

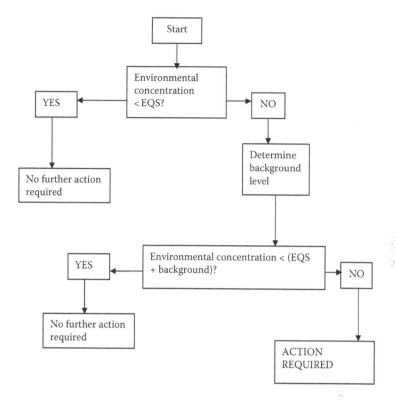

FIGURE 4.2 Summary of the added risk approach for dealing with background concentrations of metals.

phytoestrogens arising from vegetation (Safe and Gaido 1998; Shore and Shemesh 2003), and PAHs from natural combustion (Law et al. 1997). However, the principle obviously does not apply to natural substances when mobilized anthropogenically (e.g., estrogens in sewage), and the difficulty of distinguishing human from nonhuman sources will limit the scope of this approach. Further, the use of added risk for substances for which no safe threshold can be defined, such as PAHs, is scientifically and politically difficult to justify.

4.10 PROTECTION OF THE HUMAN AND WILDLIFE FOOD CHAINS FROM AQUATIC CONTAMINANTS

4.10.1 PROTECTION OF THE FOOD CHAIN — HUMANS AND WILDLIFE

4.10.1.1 Humans

For humans, the key issues relate to the demonstration that significant accumulation does not occur in a food organism. Exposure from the food organism is dependent on factors such as how much is eaten and how frequently, other sources of exposure, and the impact of preparation and cooking (e.g., the USEPA produces guidance relating to whole fish, fish fillet, cooked and uncooked fish; USEPA 2000). Guidelines listing safe levels of a number of key contaminants in food can be found in the deliberations of organizations such as the World Health Organization/Food and Agriculture Organization (WHO/FAO) Joint Expert Committee on Food Additives and Contaminants (JECFA) (e.g., mercury and methylmercury; IPCS/WHO 2000) and for pesticides the WHO/FAO Joint Meeting on Pesticides Residues (JMPR).

In carrying out a risk assessment, it is also necessary to consider by whom the food will be eaten and whether this will include high-risk groups such as weaned infants. The EU WFD only takes into account fishery products (finfish and shellfish). Data from sources such as WHO or the European Expert Committee on Food are then used to identify tolerable or acceptable intakes followed by linking these to the aquatic bioconcentration factor (BCF) for that substance. The highest reliable BCF is used, and the calculation of daily exposure is then made. This is then ideally related to a water concentration or, more probably, a sediment concentration since these substances are by definition usually highly lipophilic. The United States has a hierarchy of preference for identifying suitable BCFs (USEPA 2000). This is first a field-derived BCF, second a measured BCF, and as a last resort, use of a predicted BCF. In other countries, such as the United Kingdom, the tissue residues identified for protection of humans are used to trigger warnings followed by local action to identify the source and introduce measures to reduce inputs.

The problem of chemical intake via the aquatic food chain is gradually being eliminated by regulatory action to limit or withdraw chemicals that bioconcentrate or are biomagnified in the environment. One alert for this is a log K_{ow} of 3.0 or greater, and this is followed by specific studies on bioconcentration that take into account metabolism and depuration. For example, PAHs could theoretically bioconcentrate in vertebrate tissues but are rapidly metabolized and depurated, so bioconcentration does not occur in practice.

4.10.1.2 Wildlife

The situation regarding wildlife is different from that in humans since the issue relates largely to ingestion of complete food organisms rather than defined body parts such as muscle, and animals of course do not cook their food. For most substances, control of residue levels in wildlife cannot be reliably assumed on the basis of compliance with EQSs in water or sediment. Furthermore, it is clear that BCFs derived from continuous exposure studies can be misleading in field situations, where exposure may be brief (e.g., many modern pesticides). The requirement, therefore, is that data should be available on residues in the target species, such as fish-eating birds or mammals, but these are relatively mobile, so local conditions and geography may have an important modulating role.

The key decision is then whether the acceptable body burden in the target species is exceeded. Canada has derived Tissue Residue (i.e., body burden) Guidelines for the Protection of Wildlife Consumers of Aquatic Biota for several substances (e.g., DDT, methylmercury, PCBs), and the EU WFD permits the establishment for some substances of a body-burden EQS recalculated as an equivalent concentration in water (Lepper 2005), but there are few other approaches of this type. As indicated, the fact that most bioaccumulative substances are now under strict regulatory control reduces the potential for future problems with target species.

4.10.2 PROTECTION OF DRINKING WATER

The approach of WHO (2006) for the *Guidelines on Drinking Water Quality* was based on an international consensus of experts and stakeholders, resulting in the development of a range of guideline values for key chemical contaminants. This is the point of departure for setting standards for drinking water in many parts of the world, including the European Union. The WHO guidelines include advice on practical considerations, treatment, and analysis and on interpreting the significance of exceeding a guideline value with a transparent description of how the guideline value was derived. WHO also provides information on the management of chemicals in relation to drinking water contamination as part of a new holistic strategy for managing drinking water safety and quality from the source to the consumer. This encompasses microbial contamination (top priority) and chemical contaminants and constituents. This strategy, called Water Safety Plans, includes source water protection and control of hazards and risks throughout the system (WHO 2006).

Some countries do set environmental guidelines or standards (EGSs) for waters to be used for abstraction (USEPA 2000; for drinking water maximum contaminant levels [MCLs], see http://www.epa.gov/safewater/contaminants/index.html#mcls). However, since drinking water can and does undergo varying levels of treatment, the use of drinking water standards for treated water is particularly appropriate. The utility of standards related to abstraction in the European Union may be of less value in the future because of advances in drinking water treatment and because of the European Union's new holistic Water Safety Plan approach, although it should be noted that avoiding pollution may in some situations be cheaper than water treatment (the US approach, e.g., is focused on source water protection), and high-tech treatment may not be an option in developing countries.

4.10.3 PROTECTION OF RECREATIONAL WATER

The guidelines for recreational waters (WHO 2003) consider chemicals and microorganisms. A number of other countries also produce standards for bathing water, but these are primarily aimed at microorganisms rather than chemical substances. For toxics, except in highly unusual circumstances such as a spill of large amounts of a chemical, the only substances of significant concern with respect to recreational uses of water are considered to be the toxins from blue-green algae or marine dinoflagellates. This is primarily due to the limited exposure to other substances and organisms during recreational activities.

A guideline value for a specific group of toxins, the microcystins, is under development in Canada, and Australia has developed guidelines based on toxic algal densities (Chorus and Bartram 1999; ANZECC/ARMCANZ 2000). These guidelines trigger action to warn potential users of the risks related to a specific water body. There are no specific guideline values relating to coastal waters.

4.11   CONSIDERATION OF CARCINOGENICITY, MUTAGENICITY, AND REPRODUCTIVE TOXICITY, INCLUDING ENDOCRINE DISRUPTION

4.11.1   BACKGROUND TO CARCINOGENICITY, MUTAGENICITY, AND REPRODUCTIVE TOXICITY

The subset of environmental contaminants that may induce irreversible long-term toxicities in wildlife is an important aspect in assessing and seeking to protect surface water and sediment quality. These contaminants are often described as chemicals that are carcinogenic, mutagenic, or toxic to reproduction or endocrine disrupters ("CMR/E chemicals") (EC 2003). Prioritization schemes for such chemicals have been described for North America and Europe (Swanson et al. 1997; Baun et al. 2006), and relevant monitoring and risk assessment activities are undertaken in the United States (Claxton et al. 1998), European Union (under the Dangerous Substances Directive 76/464), and elsewhere. From a practical perspective, the working group felt that existing EQS assessments based on conventional aquatic and sediment chronic toxicity tests would usually take account of CMR/E mechanisms, particularly if the assessment is based on chronic or reproductive toxicity. This reflects the reasonable assumption that CMR/Es are a minor subset of environmental contaminants and acknowledges the increasing evidence of "practical thresholds" in toxicology (Bolt et al. 2004). Therefore, for ethical and economic reasons, only on a case-by-case basis would it be justifiable to conduct animal tests with specific CMR/E endpoints. The following text in this section aims to summarize what could be done in wildlife species, if required, to support a robust EQS in nonroutine situations.

The human and wildlife health implications of carcinogens and mutagens entering aquatic ecosystems have been studied intensively for several decades (Royal Society 1994; Belfiore and Anderson 2001; Jha 2004). For example, Claxton et al. (1998) reported that during 1994 in the United States alone more than 20 000 different facilities released 2.26 billion pounds of toxic substances into the environment, of which approximately a third were rodent carcinogens. Most of these carcinogens were discharged as components of complex mixtures (e.g., liquid effluents, airborne emissions, and solid wastes). Other scientists argued that the wide-scale focus on such synthetic chemicals needs to be balanced by consideration of natural carcinogens and mutagens (Ames and Gold 2000). In principle, cancer may occur due to either genetic mechanisms (e.g., DNA or chromosomal damage) or epigenetic or nongenotoxic mechanisms (e.g., inflammatory responses or excessively high-dose laboratory testing) (Brusick 1987; Laskin and Pendino 1995; Ames and Gold 2000).

With respect to reproductive toxicology, concerns over endocrine-disrupting chemicals have become an intensive theme of research on both human and wildlife health (Colborn et al. 1996). For the assessment and testing of estrogens and other endocrine disrupters in aquatic life, see the work of Vos et al. (2000), Hutchinson et al. (2000), and Matthiessen (2003). Importantly, however, assessment of reproductive toxicity may also involve a genetic aspect in terms of the teratogens, congenital abnormalities, and heritable gene mutation (Khera 1981; Brusick 1987; Bentley et al. 1994).

4.11.2 CARCINOGENS, MUTAGENS, AND AQUATIC ORGANISMS

4.11.2.1 Population Perspective

From a purely scientific perspective, although cancer (neoplasia) has been reported in some fish populations (Harshbarger and Clark 1990), many scientists consider that it is not a critical factor for the long-term viability of most wildlife populations given that cancer is a feature of old age, and therefore senescent or moribund individuals will generally be removed from the population by predation or other natural factors before cancer is expressed (Würgler and Kramers 1992; Jha 2004). In general, cancer (neoplasia) has been less well studied in plants and invertebrates (Gaspar 1998; Sparks 2005). It should be borne in mind, however, that there could be aesthetic and marketing implications of cancer in commercially important finfish and shellfish species (S. Feist, personal communication 2006).

On a wider scientific horizon, Kurelec (1993) proposed the "genotoxic disease syndrome" to describe the fact that genotoxic chemicals can have negative impacts on a range of important biochemical, physiological, developmental, and reproductive endpoints (the Darwinian fitness concept). Environmental stress may be exogenous (e.g., temperature, pathogens, predators) or intrinsic in origin. Such intrinsic stress could originate from an increase in homozygosity as a result of genetic drift and/ or inbreeding. These homozygous conditions, in conjunction with natural selection, will cause genetic stress that could lead to a decrease in fitness. Several mechanisms have been suggested, including an increase in homozygosity for recessive deleterious alleles and a decrease in the level of heterozygotes for overdominant loci.

Other mechanisms, such as a decrease in the level of genomic coadaptation or an increase in genomic instability, have also been suggested to play important roles in decreased fitness (Jha 2004). In this population context, genotoxins can impair the ability of aquatic organisms to develop, grow, and reproduce successfully to maintain population viability. For example, a number of studies have successfully applied this concept to aquatic annelids (Hutchinson et al. 1998), crustaceans (Atienzar and Jha 2004), echinoderms (Pagano et al. 2001), fish (Theodorakis et al. 1999), molluscs (Dixon, 1982), and plants (Atienzar et al. 2000).

4.11.2.2 Deriving PNECs for Genotoxins

Adopting the ideas proposed by Würgler and Kramers (1992) and Kurelec (1993), it is proposed that the setting of water or sediment PNEC values for genotoxins should be based on sublethal biological endpoints expressed as statistically robust ECx or NOEC values. In practice, this could include the in vivo measurement of reproduction, development, growth, or specific cytogenetic macrolesions (namely, micronuclei, aneuploidy, and chromosomal aberrations) since these are known to relate to adverse phenotypic outcomes (Brusick 1987; Jha 2004).

In vitro genetic toxicity data (e.g., Ames and umu mutagenicity assays) should not be used for deriving aquatic or sediment PNEC values but rather to guide the efficient design of in vivo tests. Genetic microlesions (i.e., measurements of DNA strand breakage or DNA repair) also should not be used for direct PNEC derivation given uncertainties regarding their relevance to adverse phenotypes in aquatic organisms (Würgler and Kramers 1992; Jha 2004). Instead, such data are probably most useful for a "mode-of-action approach" to guide the efficient design of in vivo testing for phenotypic endpoints (e.g., embryo development assays) or genetic macrolesions (e.g., aneuploidy assays) in the aquatic species of interest.

4.11.2.3   Genotoxicity Assessment Methods

There are a steadily growing number of aquatic organisms being assessed for genetic toxicity, often adopting methods originally developed for mammalian applications (e.g., Depledge 1998). While the bulk of the work reflects basic research into aquatic genetic toxicology, there are signs that regulatory agencies are interested in having internationally validated methods available. For example, ISO is developing both in vitro (V70 cell line assay) and aquatic in vivo (amphibian micronucleus assay) test methods (ISO 2006a, 2006b). It is beyond the current scope to review this broad field, but some examples are summarized in Table 4.11.

4.11.3   Reproductive Toxins, Endocrine Disrupters, and Aquatic Organisms

4.11.3.1   Population Perspective

Maintaining healthy fish and wildlife populations clearly requires that organisms' normal sexual development and reproductive functions are not impacted by chemical contaminants. As reviewed by Vos et al. (2000) and Matthiessen (2003), there is considerable evidence suggesting that the widespread occurrence of intersex fish in English rivers is probably due to estrogens. Experimental laboratory studies also show that receptor-mediated endocrine-disrupting chemicals (i.e., androgen and estrogen agonists and antagonists) can also adversely affect the sexual development and reproduction of several fish species (Hutchinson et al. 2006).

For molluscs, there is also compelling international evidence of triorganotins having caused imposex and consequential population declines, in this case probably through metabolic endocrine disruption via aromatase inhibition or other androgenic mechanisms (Schulte-Oehlmann et al. 2000; Matthiessen 2003; Duft et al. 2005; Oehlmann et al. 2007).

4.11.3.2   Reproductive Toxicity and PNEC Derivation

Based on current evidence, the derivation of EQS and PNEC values for pelagic, benthic, or sediment-dwelling fauna should generally use the standard approach of conducting chronic tests and calculating NOEC or ECx values for the sublethal endpoints (e.g., growth, development, and reproduction). It should, however, be noted that some endocrine-disrupting effects would not be detected by standard chronic tests.

An important question in this area relates to the widespread use of exposure biomarker responses for monitoring effluents and receiving waters for endocrine disrupter activity. In oviparous species, vitellogenin is the major example of such a

TABLE 4.11

Examples of the assessment of genetic microlesions and macrolesions, or genotoxin-induced phenotypic damage, in aquatic organisms

Level of biological impact	In vivo endpoint	Recommended for direct use in PNEC setting	Reference
Genetic microlesion	DNA damage — Comet assay	No	Cotelle and Férard 1999
	DNA damage — Random amplified polymorphic DNA (RAPD)	No	Atienzar and Jha 2004
Genetic macrolesion	Chromosomal structure in amphibians — micronuclei	Yes	ISO 2006a
	Chromosomal structure — chromosomal aberrations	Yes	Jha et al. 1996
	Chromosome number — aneuploidy	Yes	Dixon 1982
Phenotypic damage	Inhibition of embryo–larval development	Yes	Pagano et al. 2001
Phenotypic damage	Histopathological evidence of cancer induction	Yes	Hawkins et al. 1988

biomarker response that may be measured at the molecular, biochemical, or histological level (Sumpter and Jobling 1995). Other endocrine disrupter biomarkers include choriogenin, aromatase, and plasma sex steroid levels (Tyler et al. 1998). It has been proposed that these endocrine disruption biomarkers are best used as "signposts" to help prioritize the efficient design of field-monitoring and laboratory testing programs (e.g., species and endpoint selection) (Hutchinson et al. 2006). As the concerns over endocrine disruption extend to other modes of action, including thyroid disrupters or metabolically active agents (e.g., azole fungicides), biomarkers of absorption, distribution, metabolism, and excretion ("ADME biomarkers") will gain further scientific value for the development of efficient testing strategies to support environmental risk assessments (Figure 4.3).

Although to some extent controversial, there is evidence that some endocrine disrupters such as xenoestrogens may show nonmonotonic dose–response relationships in certain test systems, including in vivo (e.g., Weltje et al. 2005; Calabrese 2005), and this could potentially cause problems when deriving EQSs in the future.

4.11.3.3 Reproductive and Sexual Development Toxicity Assessment Tools

Briefly, there are many nationally (ASTM, Environmental Canada, and USEPA) and internationally validated methods available for assessing developmental and reproductive effects in aquatic plants and animals (ASTM, Environment Canada, ISO, and OECD). For plants, examples include the microalgal, freshwater macrophytes and marine algal growth tests (Nyholm and Källquist 1989; Wang 1990; Eklund 2005).

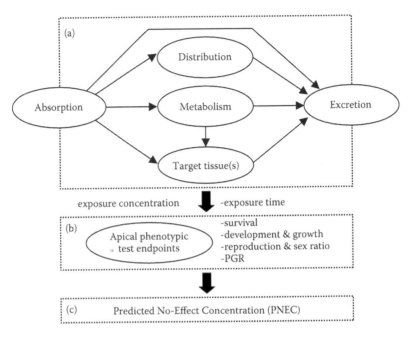

FIGURE 4.3 Principles for understanding the mechanisms and effects of endocrine disrupters or other xenobiotics in aquatic animals showing a) the ADME phase in which mechanisms are examined and can help to efficiently select test species and design chronic test methods; b) apical test endpoints for definition of adverse population effect endpoints, including population growth rate (PGR), commonly expressed as [adverse]NOEC (or NOAEC) or [adverse]EC10; and c) PNEC derivation for environmental risk assessment and EQS applications based on adverse population effect endpoints. (Adapted from Hutchinson 2007.)

For freshwater invertebrates, frequently used species are the pelagic crustacean *Daphnia magna* and the sediment-dwelling insect *Chironomus* sp., while marine crustacean test methods have used copepods and mysids. Molluscs and echinoderms are also important invertebrate species for developmental and reproductive effects assessment (EC 2003). Reproductive and developmental inhibition may be caused by both endocrine and nonendocrine modes of action; however, based on current knowledge, PNEC assessments should be based on impaired fitness parameters (e.g., reduced rates of fertility, development, or fecundity) and not on molecular or biochemical changes (Ingersoll et al. 1999; Hutchinson 2002; Barata et al. 2004).

Fish partial and full life-cycle test methods have been used successfully for many years to assess the effects of nonendocrine active chemicals (McKim 1977) and more recently to focus on high-priority endocrine disrupters (Tyler et al. 1998; Huet 2000; Länge et al. 2001). In some cases, there may also be a need to measure the potential for chronic effects on amphibians, for which developmental effect test methods are available (Devillers and Exbrayat 1992; Pickford et al. 2003). Again, in line with current knowledge, PNEC assessments should be based on impaired fitness parameters (e.g., reduced rates of fertility, development, or fecundity) and not on molecular or biochemical changes (see Figure 4.3).

4.12 VALIDATION, IMPLEMENTATION, AND REVIEW OF AQUATIC EQSs

4.12.1 VALIDATION OF AQUATIC EQSs

"Validation" can be simply described as proving that what you expect to happen actually happens, and that it happens every time. But what do we expect in EQS derivation? Do we expect that the numerical value that we estimated for the EQS is correct, or do we expect that this value represents a reliable level of protection in the field?

Different stakeholders, industry, nongovernmental organizations (NGOs), and the general public all demand EQSs that are scientifically sound and represent an adequate level of protection, notwithstanding that the level of protection required will differ among these stakeholders. Industry will generally take into account the costs that arise from taking measures to reach levels below a quality standard and may propose factors that modify the level of protection, whereas NGOs will probably emphasize the precautionary principle in setting the standards.

4.12.1.1 Validation of Correct Derivation

We agree that organizations that derive EQSs should validate (check) whether the EQS methodology has been followed correctly. The European TGD as well as the guidance for the WFD dedicate a whole section to the selection procedure for representative data (EC 2003; Lepper 2005). In other geographical areas, the procedures for selecting data that are representative and reliable are also described. A similar approach is followed in other procedures for which screening of data is essential, such as the OECD screening of High Production Volume (HPV) chemicals (OECD 2004). In several approaches, criteria according to Klimisch et al. (1997) are being applied in the data selection procedure. The application of the various assessment procedures is also described thoroughly in the above-mentioned documents. A crucial part of the whole derivation procedure seems to be correct assessment of the data and application of sufficient insight into the factors that may modify the assessment, such as pH, chemical form in which the substance is present under field circumstances, and bioaccumulation. Therefore, the assessment needs to be carried out by experts who have at least some knowledge of these modifying factors, and the outcome may be checked through peer review. Such a peer review group was described by Zabel and Cole (1999) for the United Kingdom and is also used in other countries.

4.12.1.2 Validation of EQS in the Field

How do we validate whether the EQS value is accurate in terms of reflecting the predicted impact in the real world? We suggest that validation can be achieved by inspection of field data. For example, the USEPA carried out validation of sediment EQSs in a 2-step approach: First, they conducted single-species lab experiments with spiked contaminant levels in sediments (low to high organic carbon content); second, they conducted a field validation (in Canadian experimental lakes) using a "contaminated sediment test bed." Using this method, the USEPA validated EQS approaches for a suite of metals (repeated at varying defined organic carbon levels)

(USEPA 2005b). Similar approaches were used to validate EQS methods for PAHs (USEPA 2003d).

The USEPA has also validated freshwater EQS approaches (10 compounds to show proof of principle). It deploys this validation approach each time a new method (e.g., aquatic guidelines or sediment guidelines methodology) is being developed but not for every occasion when deriving a specific substance EQS. The United States addresses bioavailability in this approach (USEPA 1986).

The goal of the validations is to ensure that concentrations at or below the EQS will not cause an adverse effect. The EQS therefore indicates the concentration below which an unacceptable adverse effect is unlikely to occur. Concentrations above the EQS may or may not cause effects depending on both the magnitude and frequency of the exceedance. In the USEPA approach, as in the aquatic life water quality criteria approach, newly generated data exceeding the standard (for that matter, all newly generated data of acceptable quality) will not replace this standard but will be added to the database from which data for estimating the standard are taken.

Similar validation is conducted by some other regulatory authorities. For example, it was shown in an Australian study that there was no effect on sensitive species at concentrations 10 times above the EQS values for metals in sediments because of the EQS derivation method in which toxicity was equally ascribed to cooccurring chemicals (leading to concerns over the ecological realism of sediment metal EQS values) (Simpson et al. 2005). In Canada, no formal validation programs of this type have been undertaken.

In summary, we are in favor of validating EQSs by inspecting field data.

4.12.2 VALIDATION PROCEDURES

Validation of the level of protection may be explored in 2 different ways. First, there should be agreement on the level of protection needed. For example, this level can be scientifically described as the 5th percentile of the SSD or the lowest NOEC divided by an AF based on laboratory experiments. But what level of protection would this represent in field situations, and what level of protection is desired in the field? It is recognized that the major input for this discussion should not come primarily from scientific fora but from policy makers. Second, a number of outstanding issues resulting from the extrapolation of laboratory studies to the field situation should be clarified. This may result in reducing the uncertainties in the derivation and in the application of EQSs in the real world. Validation with mesocosm or field data may be a good approach to validating the EQS but only as a pass-or-fail issue. It may indicate if an EQS is too high for the situation in that field site at that time, but it does not provide information on the frequency of failure or say anything about the frequency of failures that we want to accept. Thus, further exploration is needed in this direction.

Validation of EQSs is requested by stakeholders with different backgrounds. The NGOs as well as industry will request sound EQSs. For industry, predictability and stability are probably one of the main drivers. Peer review with representatives from different backgrounds is accepted in various countries, such as The Netherlands and the United Kingdom, and is mandatory in the United States (USEPA 2006). It may broaden the acceptance among stakeholders, but care should be taken that such peer review groups are science led.

4.12.3 Use of Microcosm, Mesocosm, and Field Studies for Validation Purposes

All model ecosystem studies are essentially one-off experiments, so they will inevitably require the use of evidence-based judgment and clear data-recording and audit trails. Several reviews have been published in which the outcomes of the standard test species and/or the SSD approaches are compared to threshold (NOEC) values obtained from model ecosystem studies. For example, see the work of Okkerman et al. (1993), Emans et al. (1993), Belanger (1997), Brock et al. (2000a, 2000b), and Maltby et al. (2005). These provide some evidence that the standard test species and the SSD approaches are generally protective of aquatic ecosystems. If there is uncertainty about the risk posed by a specific substance, model ecosystems as a higher-tier test system can be used to reduce this uncertainty and to calculate a refined EQS. However, it does not seem necessary, and is not envisaged, that model ecosystems should be used to validate all EQSs.

Mesocosm chronic NOEC data may also be used to set a less-stringent EQS than an EQS based purely on single-species chronic tests if the latter is challenged by a stakeholder (see Girling et al. 2000). This approach first requires evidence-based judgment on the design of a suitable mesocosm study, a stakeholder to conduct (or at least fund) the study, then further judgment about whether the mesocosm NOEC represents the definitive scientific endpoint for deriving the EQS. This will depend inter alia on information about the species diversity in the test system as compared with that in natural ponds, streams, rivers, and lakes. Mesocosm studies are also potentially valuable for the derivation of freshwater sediment EQS values and could, in principle, provide valuable corroboration for single-species laboratory sediment tests.

However, this does not answer the question of whether microcosms and mesocosms are protective of the diversity of water bodies in the field. To answer this question, it would be necessary to relate biological monitoring data (e.g., population abundances, community structure) to chemical monitoring data. Due to the fact that in the field a number of factors might affect populations and communities (e.g., habitat structure, nutrient levels, other chemicals), it is important to find data sets that allow quantification of the effect of 1 specific substance.

An example of field data being used to validate an EQS (based on single-species data) concerns the United Kingdom's development of organophosphate (OP) insecticide EQSs, which used field data on aquatic macroinvertebrate declines in otherwise unimpacted streams receiving sheep dip drainage to provide corroborative support (Lewis and Young 2000). However, note that laboratory data (Moore and Waring 1996) have demonstrated that OP concentrations even lower than those shown to be safe in mesocosm studies with invertebrates are able to damage the olfactory epithelium of male salmon parr and interfere with pheromone-induced sperm maturation.

4.12.4   Criteria for Triggering a Review of an Established EQS

There are various criteria for triggering the review of an established EQS, but those criteria will depend on the viewpoint of the stakeholder requesting revision. For example, negative feedback from the Australian minerals industry is supporting the generation of new ecotoxicology data to refine the current sediment EQS for copper (which was felt to be too high). Science-based revisions of US environmental standards may also be conducted following pressure from commercial stakeholders.

Policy changes could be a trigger to review an established EQS. The present review of aquatic EQSs in the EU member states has been instigated because of policy (or by proposed legislation). For example, revision of standards in The Netherlands is discussed within a group of policy makers from different ministries. However, within the public sector there are also different points of view. The Department for Drinking Water Supply, for example, takes a different approach from the Department for Surface Water Quality. Some technically sound and up-to-date EQS proposals have been withdrawn for wider political reasons (e.g., the Dutch standards for mineral oil in surface water and sediment). Recently, new standards derivation procedures in The Netherlands have allowed industry the possibility of deriving standards. These follow standard procedures and have to pass a peer review before being implemented. There had been several requests for revision as some pesticide industries considered the standards for their product to be too low, thus possibly hampering authorization under the Plant Protection Products (PPP) directive. Providing or generating new data may increase the EQS concentration by decreasing the AF. Generally, however, industry does not favor regular changes of standards. A summary of factors that might drive an EQS review is given in Table 4.12.

If only technical aspects are considered as triggers for EQS reviews, a few criteria emerge, such as age of the EQS, the number of data, the presence of new data, and the availability of new derivation techniques. Regular exceedance of the EQS may also be a trigger to search for new data with which to derive a possibly more robust standard.

A final issue concerns the frequency with which standards should be revised. We conclude that there is no clear-cut answer to this question and no single desirable review frequency; revision will depend on the substance and the regulatory jurisdiction. Some standards may never be revised, others every 2 years. An example of the latter is the ammonia standards in the United States, which have been revised 3 times in 10 years because of ongoing site-specific issues (e.g., pH, temperature) at sewage treatment plants. In Australia, ongoing revision was proposed but has been slow to occur. Canada does not have a fixed review period, only a general requirement to review the guidelines from time to time. In practice, revision only takes place when there is a perceived environmental problem or when Environment Canada has policy issues to address. Finally, in The Netherlands, a review period of 5 years was originally set, although in practice this has usually been much longer (up to 14 years). The issue of review frequency may not, however, be considered a pressing one given that the overwhelming majority of priority chemicals have as yet no EQS at all.

TABLE 4.12

Examples of drivers for reviewing an EQS

Examples	New scientific information	Change in regulation	Stakeholder perceptions	Reference
Metals (Australian example)	Information on toxicity	Australian sediment guidelines	Industry-funded reevaluation	Simpson et al. 2005
Ammonia (United States)	Site-specific information		NGOs, municipal wastewater treatment plant operators, and governments requested	USEPA 1999
Copper (European Union)			Voluntary risk assessment[*]	—
Lead (European Union)			Voluntary risk assessment[*]	—
Mercury (European Union)		EU Water Framework Directive		Eurochlor 2004
Hexachlorobenzene (European Union)		EU Water Framework Directive		Eurochlor 2004
Hexachlorobutadiene (European Union)		EU Water Framework Directive		Eurochlor 2004
Phosphate esters (Netherlands)	Environmental concentrations			Verbruggen et al. 2005
Metals (Netherlands)	Background concentrations and bioavailability			Van Vlaardingen et al. 2005

[*] Conducted by industrial consortia

4.13   CONCLUSIONS

1) The aim of aquatic EQSs for chemicals is to protect the structure and functioning of aquatic ecosystems and to protect the food chain leading to terrestrial organisms (including humans). They may also be used to protect raw drinking water quality and recreational uses of water.

2) It will usually be necessary to derive several aquatic EQSs for a given substance to meet different protection goals.

3) EQS derivation procedures should maintain a balance between the need for prescription (to achieve consistency) and the need for evidence-based judgment (to deal with nonstandard circumstances).

4) Criteria for prioritizing substances for EQS derivation are many and various, including scientific, policy, and stakeholder drivers.

5) There is no consensus about the minimum amount of data required for deriving aquatic quality standards, but the use of very small data sets (e.g., toxicity data on 1 alga, 1 crustacean, and 1 fish species) is likely to result in unreliable PNECs.

6) Substances generally require aquatic standards protective against both short-term and long-term exposure, often expressed as a not-to-be-exceeded and a long-term concentration (e.g., MAC-EQSs and AA-EQSs).

7) While acute toxicity data are needed for deriving MAC-EQSs, it is preferable to use chronic toxicity data for deriving AA-EQSs. For this purpose, algal toxicity tests should be regarded as providing chronic data as they represent several generations of population growth.

8) Aquatic toxicity data from which EQSs are derived must have been validated according to an agreed procedure, especially if they derive from nonstandard tests. Furthermore, it is important to ensure that the derivation procedures themselves, and all decisions resulting from evidence-based judgment, are properly validated and recorded.

9) While aquatic EQSs can be derived solely using AFs applied to data from test species, it is considered more reliable to use SSDs based on larger chronic data sets (if available), and the use of good-quality model ecosystem data (again, if available) is probably the most reliable approach for deriving long-term EQSs, particularly if exposure in the tests was maintained. Mesocosms in which exposure was transient may also be used for deriving short-term EQSs.

10) The AFs, whose size should be designed to reflect expected uncertainties in the data, should generally be used when deriving aquatic EQS values from either laboratory or model ecosystem data. It may also be appropriate to apply small AFs to the outcome of SSDs.

11) The success and accuracy with which MAC- and AA-EQSs are used will depend crucially on the design of the aquatic monitoring programs that provide the data on chemical concentrations in waters and sediments, which in turn should depend on the properties of the chemical and its discharge pattern.

12) Aquatic EQSs should not generally be used in isolation but as 1 of several lines of evidence about the likely impact of a chemical on a water body.

13) Exceedance of an aquatic EQS may result in immediate regulatory action but should often lead to further chemical and/or biological investigations.

14) There is no "systematic" bias in chemical sensitivity between freshwater and marine species, but use of freshwater data to support the derivation of marine EQSs should be conducted with caution on a case-by-case basis. Marine EQSs based on "read-across" with freshwater data should be regarded as tentative rather than definitive. Furthermore, the somewhat higher biodiversity in the marine ecosystem as a whole should not "automatically" result in the use of higher AFs when deriving marine EQSs.

15) Methods for calculating sediment EQSs from water-phase EQSs (e.g., EP), for predicting toxicity based on biomarker responses or for predicting toxicity based on chemical structure, should be treated with caution and should generally result in tentative rather than definitive EQS values.

16) There is no consensus on the appropriateness of using background concentrations as the starting point for assessing EQS compliance. The jurisdictions represented at the workshop have a variety of approaches to addressing "natural substances" (e.g., metals). In the European Union, aquatic EQSs for natural substances such as metals may need to be modified locally to take account of background concentrations in waters and sediments, provided that the environmental concentration exceeds the EQS. Action should only be taken if the measured environmental concentration exceeds the background concentration for the area in question when added to the EQS concentration. In the United States, the EQS derivation process for metals (and many other substances) is designed to be site specific and thus addresses the unique water, sediment, and species of each at the time of derivation. An "added risk approach" would be duplicative. Action could be taken whenever the measured environmental concentration exceeds the site-specific value.

17) The aquatic food chain to humans and wildlife should be protected by calculating expected tissue concentrations that are predicted to result from ingestion of aquatic organisms (based on measured BCFs, possible biomagnification, and frequency of ingestion) and comparing these with toxicity data (acceptable daily intake) for birds, mammals, and so on. In the case of humans, this process must be adjusted to account for losses during food preparation.

18) EQSs may be derived from internationally agreed acceptable daily intake values and applied to raw waters for the protection of humans and animals from their ingestion, but it is often sufficient to apply these values to treated water if the treatment technology is adequate. Drinking water quality is best protected through the use of holistic Water Safety Plans that include source water protection as well as controlling hazards and risks throughout the water supply system.

19) EQSs for protecting recreational uses of water can be aimed at synthetic chemicals (especially in acute spill situations) but usually are only relevant for microorganisms and algal toxins.

20) Substances that are carcinogenic, mutagenic, or reproductively toxic (i.e., CMRs), for example, some endocrine disrupters, may pose special problems for derivation of aquatic EQSs (e.g., lack of internationally agreed tests in some cases; difficulties with prediction of safe concentrations), but use of special tests for these properties is only justified for a small subset of chemicals that meet clear criteria. Furthermore, EQSs for these substances should not be derived directly from in vitro data or from biomarkers of exposure but from in vivo tests alone.

21) It is desirable to assess whether a proposed aquatic EQS is accurate in terms of the known impacts of the chemical in the real world. This can be done by comparing the EQS with suitable model ecosystem data (if these were not used in its derivation) or better still by inspection of appropriate field data, particularly from pollution gradients for which the presence of any confounding factors (e.g., other chemicals) is well understood.

4.14　RECOMMENDATIONS

It became clear during discussions that the current practice by which many countries conduct water and sediment EQS derivations on the same substance, often by somewhat different methods, is highly inefficient and wasteful of resources. To improve this situation, we recommend that the advantages and disadvantages of international sharing of EQS data and derivation strategies should be examined. Drivers for this include the following considerations:

- Even after more than 30 years of work, most countries have developed less than 50 EQSs.
- Development of a single EQS requires at least 2 to 3 years and can cost US$50 to 150K or more, depending on data availability, levels of uncertainty that must be resolved, and any economic or social controversies about the substance.
- Most countries have similar priority substances.
- Duplication of work is wasteful; there is great potential for collaboration.
- Most EQS derivation procedures are fairly similar, so there is potential for international harmonization.
- Pollution often straddles national borders (e.g., long-range transport of air pollutants [LRTAP]), rivers flowing through several countries).
- Industry and trade are multinational (i.e., sources of pollution are international).

As a first step, it would be useful if access to data on specific substances could be made simpler and a consensus reached about which data are sufficiently reliable for EQS derivation. A long-term goal would be the derivation of internationally agreed, science-based benchmarks that could be the basis (starting point) for national EQS (or criteria, guideline, objective) derivation programs. This recognizes that different jurisdictions will have different political priorities, but that the resulting standards should still have a common scientific basis.

We therefore recommend that SETAC should take steps toward attaining this goal by setting up an environmental standards study group with the explicit remit of developing harmonized procedures for derivation of environmental benchmarks.

NOTES

1. Effects classes aim to facilitate an objective interpretation of model ecosystem experiments performed for regulatory purposes. Effect Class 1 = effects could not be demonstrated; Class 2 = "slight effects."
2. Superfund sites are the United State's worst toxic waste sites and are regulated under the Comprehensive Environmental Response, Compensation, and Liability Act (CERCLA).

REFERENCES

Aldenberg T, Jaworska JS. 2000. Uncertainty of the hazardous concentration and fraction affected for normal species sensitivity distributions. Ecotox Environ Saf 46:1–18.

Allan IJ, Vrana B, Greenwood R, Mills GA, Roig B, Gonzalez C. 2006. A "toolbox" for biological and chemical monitoring requirements for the European Union's Water Framework Directive. Talanta 69:302–322.

Ames BN, Gold LS. 2000. Paracelsus to parascience: the environmental cancer distraction. Mutat Res 447:1–13.

[ANZECC/ARMCANZ] Australia and New Zealand Environment and Conservation Council/ Agricultural and Resource Management Council of Australia and New Zealand. 2000. Australian and New Zealand guidelines for fresh and marine water quality. Canberra (Australia).

Atienzar FA, Cordi B, Donkin ME, Evenden AJ, Jha AN, Depledge MH. 2000. Comparison of ultraviolet-induced genotoxicity detected by random amplified polymorphic DNA with chlorophyll fluorescence and growth in a marine macroalgae, *Palmaria palmata*. Aquat Toxicol 50:1–12.

Atienzar FA, Jha AN. 2004, The random amplified polymorphic DNA (RAPD) assay to determine DNA alterations, repair and transgenerational effects in B(a)P exposed Daphnia magna. Mutat Res 552:125–140.

Babut M, Bonnet C, Bray M, Flammarion P, Garric J, Golaszewski G. 2003. Developing environmental quality standards for various pesticides and priority contaminants for French freshwaters. J Environ Manage 69:139–147.

Barata C, Porte C, Baird DJ. 2004. Experimental designs to assess endocrine disrupting effects in invertebrates — a review. Ecotoxicology 13:511–517.

Batley GE, Stahl RG, Babut MP, Bott TL, Clark JR, Field LJ, Ho K, Mount DR, Swartz RC, Tessier A. 2005. The scientific underpinnings of sediment quality guidelines. In: Wenning R, Batley G, Ingersoll C, Moore D, editors. Use of sediment quality guidelines and related tools for the assessment of contaminated sediments. Pensacola (FL): SETAC Press, p. 39–120.

Bau M, Dulski P. 1996. Anthropogenic origin of positive gadolinium anomalies in river waters. Earth Planet Sci Lett 143:245–255.

Baun A, Eriksson E, Ledin A, Mikkelsen PS. 2006. A methodology for ranking and hazard identification of xenobiotic organic compounds in urban stormwater. Sci Tot Environ 370:29–38.

Belanger SE. 1997. Literature review and analysis of biological complexity in model stream eco-systems: influence of size and experimental design. Ecotox Environ Saf 52:150–171.

Belanger SE, Meiers EM, Bausch RG. 1995. Direct and indirect ecotoxicological effects of alkyl sulfate and alkyl ethoxysulfate on macroinvertebrates in stream mesocosms. Aquat Toxicol 33:65–87.

Belfiore NM, Anderson SL. 2001. Effects of contaminants on genetic patterns in aquatic organisms: a review. Mutat Res 489:97–122.

Bentley KS, Sarrif AM, Cimino MC, Auletta AE. 1994. Assessing the risk of heritable gene mutation in mammals: *Drosophila* sex-linked recessive lethal test and tests measuring DNA damage and repair in mammalian germ cells. Environ Mol Mutagen 23:3–11.

Bolt HM, Foth H, Hengstler JG, Degen GH. 2004. Carcinogenicity categorization of chemicals — new aspects to be considered in a European perspective. Toxicol Lett 151:29–41.

Brock TCM, Arts GHP, Maltby L, Van den Brink PJ. 2006. Aquatic risks of pesticides, ecological protection goals, and common aims in European Union legislation. Integ Environ Assess Manag 2:20–46.

Brock TCM, Lahr J, Van den Brink PJ. 2000a. Ecological risks of pesticides in freshwater ecosystems. Part 1: Herbicides. Wageningen (NL): Alterra Report 088. 127 p.

Brock TCM, Van Wijngaarden RPA, Van Geest GJ. 2000b. Ecological risks of pesticides in fresh-water ecosystems. Part 2: insecticides. Wageningen (NL): Alterra Report 089. 142 p.

Brusick D. 1987. Principles of genetic toxicology. 2nd ed. New York: Plenum Press. 284 p.

Bryan GW, Gibbs PE, Hummerstone LG, Burt GR. 1987. Copper, zinc, and organotin as long-term factors governing the distribution of organisms in the Fal Estuary in southwest England. Estuaries 10:208–219.

Calabrese EJ. 2005. Paradigm lost, paradigm found: the re-emergence of hormesis as a fundamental dose-response model in the toxicological sciences. Environ Pollut 138:378–411.

Campbell PJ, Arnold DJS, Brock TCM, Grandy NJ, Heger W, Heimbach F, Maund SJ, Streloke M. 1999. Guidance document on higher-tier aquatic risk assessment for pesticides (HARAP). Brussels: SETAC-Europe. 179 p.

Caurant F, Aubail A, Lahaye V, Van Canneyt O, Rogan E, López A, Addink M, Churlaud C, Robert M, Bustamante P. 2006. Lead contamination of small cetaceans in European waters — the use of stable isotopes for identifying the sources of lead exposure. Mar Environ Res 62:131–148.

[CCME] Canadian Council of Ministers of the Environment. 1991. Appendix IX. A protocol for the derivation of water quality guidelines for the protection of aquatic life (April 1991). In: Canadian water quality guidelines, Canadian Council of Resource and Environment Ministers, 1987. Prepared for the Taskforce on Water Quality Guidelines. [Updated and reprinted with minor revisions and editorial changes in Canadian environmental quality guidelines, Chapter 4, Winnipeg (Manitoba). 1999.] 10 p.

[CCME] Canadian Council of Ministers of the Environment. 1999a. Protocol for the derivation of Canadian sediment quality guidelines for the protection of aquatic life. Canadian Council of Ministers of the Environment 1995. CCME EPC-98E. 35 p.

[CCME] Canadian Council of Ministers of the Environment. 1999b. Protocol for the derivation of Canadian tissue residue guidelines for the protection of wildlife that consumes aquatic biota. Winnipeg (Manitoba). 1998. 18 p.

[CCME] Canadian Council of Ministers of the Environment. 2000. Canadian water quality guidelines for the protection of aquatic life: Ammonia. In: Canadian environmental quality guidelines, 1999. Winnipeg (Manitoba).

[CCME] Canadian Council of Ministers of the Environment. 2002. Canadian water quality guidelines for the protection of aquatic life: inorganic fluorides. In: Canadian environmental quality guidelines, 1999. Winnipeg (Manitoba).

Chapman PM, Caldwell RS, Chapman PF. 1996. A warning: NOECs are inappropriate for regulatory use. Environ Toxicol Chem 15:77–79.

Chorus I, Bartram J. 1999. Toxic cyanobacteria in water. A guide to their public health consequences, monitoring and management. London (UK): E and FN Spon.

Claxton LD, Houk VS, Hughes TJ. 1998. Genotoxicity of industrial wastes and effluents. Mutat Res 410:237–243.

Clean Water Act (Federal Water Pollution Control Act) (as amended through P.L. 107-303, November 27, 2002). Available from: http://www.epa.gov/lawsregs/laws/cwa.html accessed 23 June 2009.

Colborn T, Dumanoski D, Myers JP. 1996. Our stolen future: are we threatening our fertility, intelligence, and survival? A scientific detective story. New York: Dutton.

Comber MHI, Walker JD, Watts C, Hermens J. 2003. Quantitative structure–activity relationships for predicting potential ecological hazard of organic chemicals for use in regulatory risk assessments. Environ Toxicol Chem 22:1822–1828.

Cotelle S, Férard JF. 1999. Comet assay in genetic ecotoxicology. A review. Environ Mol Mutagen 34:246–255.

Crommentuijn T, Polder M, Sijm D, de Bruijn J, Van de Plassche E. 2000. Evaluation of the Dutch environmental risk limits for metals by application of the added risk approach. Environ Toxicol Chem 19:1692–1701.

Depledge MJ. 1998. The ecotoxicological significance of genotoxicity in marine invertebrates. Mutat Res 399:109–122.

Devillers J, Exbrayat JM. 1992. Ecotoxicity of chemicals to amphibians. Lyon (France): Gordan and Breach Science.

de Wolf W, Siebel-Sauer A, Lecloux A, Koch V, Holt M, Feijtel T, Comber M, Boeije G. 2005. Mode of action and aquatic exposure thresholds of no concern. Environ Toxicol Chem 24:479–485.

Di Toro DM, Allen HE, Bergman HL, Meyer JS, Paquin PR, Santore RC. 2001. Biotic ligand model of the acute toxicity of metals. 1. Technical basis. Environ Toxicol Chem 20:2383–2396.

Di Toro DM, Mahony JD, Hansen DJ, Scott KJ, Hicks MB, Mayr SM, Redmond MS. 1990. Toxicity of cadmium in sediments: the role of acid volatile sulfide. Environ Toxicol Chem 9:1487–1502.

Di Toro DM, McGrath JA, Hansen DJ, Berry WJ, Paquin PR, Mathew R, Wu KB, Santore RC. 2005. Predicting sediment metal toxicity using a sediment biotic ligand model: methodology and initial application. Environ Toxicol Chem 24:2410–2427.

Di Toro DM, Zarba CS, Hansen DJ, Berry WJ, Swartz RC, Cowan CE, Pavlou SP, Allen HE, Thomas NA, Paquin PR. 1991. Technical basis for establishing sediment quality criteria for nonionic organic chemicals using equilibrium partitioning. Environ Toxicol Chem 10:1541–1583.

Dixon DR. 1982. Aneuploidy in mussel embryos (*Mytilus edulis* L.) originating from a polluted dock. Mar Biol Lett 3:155–161.

Duboudin C, Ciffroy P, Magaud H. 2004. Effects of data manipulation and statistical methods on species sensitivity distributions. Environ Toxicol Chem 23:489–499.

Duft M Schulte-Oehlmann U, Tillmann M, Weltje L, Oehlmann J. 2005. Biological impact of organotin compounds on molluscs in marine and freshwater ecosystems. Coastal Mar Sci 29:95–110.

Durda JL, Preziosi DV. 2000. Data quality evaluation of toxicological studies used to derive ecotoxicological benchmarks. Human Ecol Risk Assess 6:747–765.

[EC] European Commission. 1999. Study on the prioritisation of substances dangerous to the aquatic environment. I. Revised proposal for a list of priority substances in the context of the Water Framework Directive (COMPPS procedure). Brussels: EC report no. CR-24-99-510-EN-C. 262 p.

[EC] European Commission. 2003. Technical guidance document in support of Commission Directive 93/67/EEC on risk assessment for new notified substances, and Commission Regulation (EC) No 1488/94 on risk assessment for existing substances, and Directive 98/8/EC of the European Parliament and of the Council concerning the placing of biocidal products on the market. 2nd ed. Luxembourg: Office for Official Publications of the European Communities. 1009 p.

[ECETOC] European Center for Ecotoxicology and Toxicology of Chemicals. 1997. The value of aquatic model ecosystem studies in ecotoxicology. ECETOC technical report no. TR 073.

[ECETOC] European Center for Ecotoxicology and Toxicology of Chemicals. 1998. QSARs in the assessment of the environmental fate and effects of chemicals. ECETOC technical report no. TR 074, European Centre for Ecotoxicology and Toxicology of Chemicals. Brussels, Belgium.

Eklund B. 2005. Development of a growth inhibition test with the marine and brackish water red alga *Ceramium tenuicorne*. Mar Pollut Bull 50:921–930.

Emans HJB, Van De Plassche EJ, Canton JH, Okkerman PC, Sparenburg PM. 1993. Validation of some extrapolation methods used for effect assessment. Environ Toxicol Chem 12:2139–2154.

Eurochlor. 2004. Chlorine industry review 2004–2005. Brussels (Belgium): Eurochlor. 29 p.

Forbes VE, Calow P. 2002. Species sensitivity distributions: a critical appraisal. Human Ecol Risk Assess 8:473–492.

Fraunhofer Institute for Molecular Biology and Applied Ecology 1999. Revised proposal for a list of priority substances in the context of the water framework directive (COMMPS procedure). Declaration ref.: 98/788/3040/DEB/E1. Final report [pdf] 97 p. Available from: <http://ec.europa.eu/environment/water/water-framework/preparation_priority_list.htm>. Accessed 23 June 2009.

Gaspar T. 1998. Plants can get cancer. Plant Physiol Biochem 36:203–204.

G-BASE. 2009. Geochemical baseline survey of the environment. British Geological Survey. Available from: <http://www.bgs.ac.uk/gbase/home.html>. Accessed 23 June 2009.

Girling AE, Tattersfield L, Mitchell GC, Crossland NO, Pascoe D, Blockwell SJ, Maund SJ, Taylor EJ, Wenzel A, Janssen CR, Jüttner I. 2000. Derivation of predicted no-effect concentrations for lindane, 3,4-dichloroaniline, atrazine, and copper. Ecotoxic Environ Saf 45:148–176.

Government of Canada. 2006. Canadian environmental sustainability indicators 2006 report. Environment Canada. Ottawa (Ontario).

Hamon RE, McLaughlin MJ, Gilkes RJ, Rate AW, Zarcinas B, Robertson A, Cozens G, Radford N, Bettenay L. 2004. Geochemical indices allow estimation of heavy metal background concentrations in soils. Global Biogeochem Cycles 18:GB1014.

Harshbarger JC, Clark JB. 1990. Epizootiology of neoplasms in bony fish of North America. Sci Total Environ 94:1–32.

Hawkins WE, Overstreet RM., Walker WW. 1988. Carcinogenicity tests with small fish species. Aquat Toxicol 11:113–128.

Huet MC. 2000. OECD activity on endocrine disrupters. Test guidelines development. Ecotoxicology 9:77–84.

Hutchinson TH. 2002. Reproductive and developmental effects of endocrine disrupters in invertebrates: in vitro and in vivo approaches. Toxicol Lett 131:75–81.

Hutchinson TH. 2007. Small is useful in endocrine disrupter assessment — four key recommendations for aquatic invertebrate research. Ecotoxicology 16:231–238.

Hutchinson TH, Ankley GT, Segner H, Tyler CR. 2006. Screening and testing for endocrine disruption in fish — biomarkers as "signposts not traffic lights" in risk assessment. Environ Health Perspect 114:106–114.

Hutchinson TH, Brown R, Brugger KE, Campbell PM, Holt M, Länge R, McCahon P, Tattersfield LJ, van Egmond R. 2000. Ecological risk assessment of endocrine disruptors. Environ Health Perspect 108:1007–1014.

Hutchinson TH, Jha AN, Mackay JM, Elliott BM, Dixon DR. 1998. Evaluation of the genotoxicity of disinfected sewage effluent using the marine worm *Platynereis dumerilii* (Polychaeta: Nereidae). Mutat Res 399:97–108.

Ingersoll CG, Hutchinson TH, Crane M, Dodson S, DeWitt T, Gies A, Huet M-C, McKenney CL, Oberdörster E, Pascoe D, Versteeg DJ, Warwick O. 1999. Laboratory toxicity tests for evaluating potential effects of endocrine disrupting compounds. In: DeFur P, Crane M, Ingersoll C, Tattersfield L, editors. Endocrine disruption in invertebrates: endocrinology, testing and assessment. Proceedings of the EDIETA Workshop 12–15 December 1998. Noordwijkerhont (NL): SETAC Technical Publications Series ISBN 1-880611-27-9, p 107–197.

IPCS/WHO. 2000. Methyl mercury. WHO Food Additives Series No. 44. Prepared for 53rd meeting of the Joint FAO/WHO Expert Committee on Food Additives (JECFA), World Health Organization, Geneva, Switzerland.

[ISO] International Organization for Standardization. 2006a. Water quality — evaluation of genotoxicity by measurement of micronuclei — Part 1: evaluation of genotoxicity using amphibian larvae. ISO/FDIS 21427-1:2006(E).

[ISO] International Organization for Standardization. 2006b. Water quality — evaluation of genotoxicity by measurement of the induction of micronuclei — Part 2: mixed population method using the cell line V79. ISO/FDIS 21427-2:2006(E).

Jha AN. 2004. Genotoxicological studies in aquatic organisms: an overview. Mutat Res 552:1–17.

Jha AN, Hutchinson TH, Mackay JM, Elliott BM, Dixon DR. 1996. Development of an in vivo marine genotoxicity test with *Platynereis dumerilii* (Polychaeta: Nereidae). Mutat Res 359:141–150.

Khera AS. 1981. Common foetal aberrations and their teratogenic significance: a review. Fund App Toxicol 1:13–18.

Klimisch HJ, Andreae E, Tillmann U. 1997. A systematic approach for evaluating the quality of experimental and ecotoxicological data. Reg Tox Pharm 25:1–5.

Kurelec B. 1993. The genotoxic disease syndrome. Mar Environ Res 35:341–348.

Länge R, Hutchinson TH, Croudace CP, Siegmund F, Schweinfurth H, Hampe P, Panter GH, Sumpter JP. 2001. Effects of the synthetic oestrogen 17α-ethinylestradiol over the life-cycle of the fathead minnow (*Pimephales promelas*). Environ Toxicol Chem 20:1216–1227.

Laskin DL, Pendino KJ. 1995. Macrophages and inflammatory mediators in tissue injury. Annu Rev Pharmacol Toxicol 35:655–677.

Law RJ, Dawes VJ, Woodhead RJ, Matthiessen P. 1997. Polycyclic aromatic hydrocarbons (PAH) in seawater around England and Wales. Mar Pollut Bull 34:306–322.

Lepper P. 2005. Manual on the methodological framework to derive environmental quality standards for priority substances in accordance with Article 16 of the Water Framework Directive (2000/60/EC). Fraunhofer Institute report. Fraunhofer Institute for Molecular Biology and Applied Ecology, Schmallenberg, Germany. 47 p.

Leung KMY, Morritt D, Wheeler JR, Whitehouse P, Sorokin N. 2001. Can saltwater toxicity be predicted from freshwater data? Mar Pollut Bull 42:1007–1013.

Lewis S, Young W. 2000. Proposed environmental quality standards for organophosphate sheep dip chemicals in water. Bristol (UK): Environment Agency. R&D technical report no. P128. 76 p.

Maltby L, Blake N, Brock TCM, van den Brink PJ. 2005. Insecticide species sensitivity distributions: importance of test species selection and relevance to aquatic ecosystems. Environ Toxicol Chem 24:379–388.

Markich SJ, Batley GE, Stauber JL, Rogers NJ, Apte SC, Hyne RV, Bowles KC, Wilde KL, Creighton NM. 2005. Hardness corrections for copper are inappropriate for protecting sensitive freshwater biota. Chemosphere 60:1–8.

Matthiessen P. 2003. Historical perspective on endocrine disruption in wildlife. Pure Appl Chem 75:2197–2206.

McKim JM. 1977. Evaluation of tests with early life-stages of fish for predicting long term toxicity. J Fish Res Board Can 34:1134–1154.

Means JC. 1995. Influence of salinity upon sediment-water partitioning of aromatic hydrocarbons. Mar Chem 51:3–16.

Moore A, Waring CP. 1996. Sublethal effects of the pesticide diazinon on olfactory function in mature male Atlantic salmon parr. J Fish Biol 48:758–775.

Moss B. 1988. Ecology of fresh waters. 2nd ed. Oxford (UK): Blackwell Scientific.

Nyholm N, Källquist T. 1989. Methods for growth inhibition tests with freshwater algae. Environ Toxicol Chem 8:689–703.

[OECD] Organization for Economic Cooperation and Development. 1984. Activated sludge, respiration inhibition test, test guideline 209, adopted 4 April 1984.

[OECD] Organization for Economic Cooperation and Development. 2004. Manual for investigation of HPV chemicals OECD Secretariat, September 2004. Available from: http://www.oecd.org/document/7/0,3343,en_2649_34379_1947463_1_1_1_1,00.html.

[OECD] Organization for Economic Cooperation and Development. 2006a. Current approaches in the statistical analysis of ecotoxicological data: a guidance to application. Paris (France): Environment Directorate OECD. OECD Environment Health and Safety Publications. Series on testing and assessment, no. 54. 147 p.

[OECD] Organization for Economic Cooperation and Development. 2006b. Guidance document on simulated freshwater lentic field tests (outdoor microcosms and mesocosms). Paris (France): Environment Directorate, Organisation for Economic Co-operation and Development. Series on testing and assessment, no. 53, ENV/JM/MONO(2006)17. 37 p.

Oehlmann J, di Benedetto P, Tillmann M, Duft M, Oetken M, Schulte-Oehlmann U. 2007. Endocrine disruption in prosobranch molluscs: evidence and ecological relevance. Ecotoxicology 16:29–43.

Okkerman PC, van der Plassche EJ, Emans HJB, Canton JH. 1993. Validation of some extrapolation methods with toxicity data derived from multiple species experiments. Ecotox Environ Saf 25:341–359.

OSPAR Commission. 1997. Agreed background reference concentrations for contaminants in seawater, biota and sediment. London: Oslo and Paris Commission. Paper no. OSPAR 97/15/1, Annex 5.

Östlund P, Sternbeck J. 2001. Total lead and stable lead isotopes in Stockholm sediments. Water Air Soil Pollut 1:229–239.

Oudin LC, Maupas D. 2003. Système d'évaluation de la Qualité des Cours d'eau — rapport de présentation (version 2). Ministère de l'Ecologie et du Développement Durable, Agences de l'eau: French Ministry of the Environment, Paris France, 106 pp.

Pagano G, de Baise A, Deeva IB, Degan P, Doronin YK, Iaccarino M, Ora, R, Trieff NM, Warnau M, Korkina LG. 2001. The role of oxidative stress in developmental and reproductive toxicity of tamoxifen. Life Sci 68:1735–1749.

Paquin PR, Gorsuch JW, Apte SC, Bowles KC, Batley GE, Campbell PGC, Delos C, Di Toro DM, Dwyer RL, Galvez F, Gensemer RW, Goss GG, Hogstrand C, Janssen CR, McGeer JC, Naddy RB, Playle RC, Santore RC, Schneider U, Stubblefield WA, Wood CM, Wu KB. 2002. The biotic ligand model: a historical overview. Comp Biochem Physiol 133C:3–36.

Peters C, Ahlf W, von Lochow HEC, Ratte HT. 2006. How should we deal with the uncertainty in the extrapolation of the sensitivity of marine organisms to narcotics? Critical review of the state of the art of marine risk assessment and recommendations for future research. Mar Sens CEFIC LRI-ECO3A-TUHH-0407. Final Report. 40 p. Available from: European Chemical Industry Council, Brussels, Belgium.

Pickford DB, Hetheridge MJ, Caunter JE, Tilghman Hall A, Hutchinson TH. 2003. Assessing chronic toxicity of bisphenol A to larvae of the African clawed frog (*Xenopus laevis*) in a flow-through exposure system. Chemosphere 53:223–235.

Posthuma L, Suter GW, Traas TP. 2002. Species sensitivity distributions in ecotoxicology. Boca Raton (FL): Lewis.

Qian SS, Lyons RE. 2006. Characterization of background concentrations of contaminants using a mixture of normal distributions. Environ Sci Technol 40:6021–6025.

Royal Society. 1994. Carcinogenesis in the marine environment, pollutant control priorities in the aquatic environment: scientific guidelines for management, report of a Royal Society study group London: Royal Society, p. 55–80.

Rüdel H. 2003. Case study: bioavailability of tin and tin compounds. Ecotox Environ Saf 56:180–189.

Russell FS, Yonge CM. 1928. The seas. London: Frederick Warne and Co. Ltd.

Safe SH, Gaido K. 1998. Phytoestrogens and anthropogenic estrogenic compounds. Environ Toxicol Chem 17:119–126.

[SANCO] Santé des Consommateurs. 2002. Guidance document on aquatic ecotoxicology in the context of the Directive 91/414/EEC. Brussels: European Commission, Health and Consumer Protection Directorate-General. SANCO/3268/2001 rev. 4 (final).

Schulte-Oehlmann U, Watermann B, Tillmann M, Scherf S, Markert B, Oehlmann J. 2000. Effects of endocrine disruptors on prosobranch snails (Mollusca: Gastropoda) in the laboratory. Part II: triphenyltin as a xeno-androgen. Ecotoxicology 9:399–412.

Shao Q. 2000. Estimation of hazardous concentrations based on NOEC toxicity data: an alternative approach. Environmetrics 11:583–595.

Shore LS, Shemesh M. 2003. Naturally produced steroid hormones and their release into the environment. Pure Appl Chem 75:1859–1871.

Simpson SL, Batley GE. 2007. Predicting metal toxicity in sediments: a critique of current approaches. Integ Environ Assess Manag 3:3–16.

Simpson SL, Batley GE, Chariton AA, Stauber JL, King CK, Chapman JC, Hyne RV, Gale SA, Roach AC, Maher WA. 2005. Handbook for sediment quality assessment. Bangor (NSW, Australia): CSIRO. 117 p. Available from: www.clw.csiro.au/cecr. Accessed 23 June 2009.

Solomon KR, Baker DB, Richards RP, Dixon KR, Klaine SJ, La Point TW, Kendall RJ, Weisskopf CP, Giddings JM, Giesy JP, Hall LW Jr, Williams WM. 1996. Ecological risk assessment of atrazine in North American surface waters. Environ Toxicol Chem 15:31–76.

Solomon KR, Giddings JM, Maund SJ. 2001. Probabilistic risk assessment of cotton pyrethroids in aquatic ecosystems: 1. distributional analyses of laboratory aquatic toxicity data. Environ Toxicol Chem 20:652–659.

Sparks AK. 2005. Observations on the history of non-insect invertebrate pathology from the perspective of a participant. J Invert Pathol 89:67–77.

Stephan CE, Mount DI, Hansen DJ, Gentile JH, Chapman GA, Brungs WA. 1985. Guidelines for deriving numerical national water quality criteria for the protection of aquatic organisms and their uses. Washington (DC): United States Environmental Protection Agency. EPA-822-R85100.

Sumpter JP, Jobling S. 1995. Vitellogenesis as a biomarker for estrogenic contamination of the aquatic environment. Environ Health Perspect 103:173–178.

Sunda W, Guillard RRL. 1976. The relationship between cupric ion activity and the toxicity of copper to phytoplankton. J Mar Res 34:511–529.

Swanson MB, Davis GA, Kincaid LE, Schultz TW, Bartmess JE. 1997. A screening method for ranking and scoring chemicals by potential human health and environmental impacts, Environ Toxicol Chem 16:372–383.

Tait RV. 1978. Elements of marine ecology. London: Butterworths.

Theodorakis CW, Elbl T, Shugart LR. 1999. Genetic ecotoxicology. Part IV. Survival and DNA strand breakage is dependent on genotype in radionuclide-exposed mosquitofish. Aquat Toxicol 45:279–291.

Tyler CR, Jobling S, Sumpter JP. 1998. Endocrine disruption in wildlife: a critical review of the evidence. Crit Rev Toxicol 28:319–361.

[USEPA] US Environmental Protection Agency. 1986. Sediment quality criteria methodology validation: calculation of screening level concentrations from field data. Washington: US Environmental Protection Agency Report No EPA822R86101.

[USEPA] US Environmental Protection Agency. 1989. Risk assessment guidance for superfund. I.: Human health evaluation manual (Part A). Technical report EPA 540-1-89-002 Washington (DC): Office of Emergency and Remedial Response, USEPA.

[USEPA] US Environmental Protection Agency. 1991. Technical support document for water quality-based toxics control. Washington: EPA/505/2-90-001. March 1991.

[USEPA] US Environmental Protection Agency. 1994. Water quality standards handbook. 2nd ed. Washington: EPA-823-B-94-005. August 1994.

[USEPA] US Environmental Protection Agency. 1999. 1999 update of ambient water quality criteria for ammonia. Washington: EPA-822-R-99-014. December 1999.

[USEPA] US Environmental Protection Agency. 2000. Methodology for deriving ambient water quality criteria for the protection of human health. Washington: EPA-822-B-00-004. October 2004.

[USEPA] US Environmental Protection Agency. 2002. Guidance for comparing background and chemical concentrations in soil for CERCLA sites. Washington: Technical report EPA 540-R-01-003, OSWER9285.7-41.

[USEPA] US Environmental Protection Agency. 2003a. Ambient aquatic life water quality criteria for atrazine — revised draft. Washington: USEPA-822-R-03-023. October 2003.

[USEPA] US Environmental Protection Agency. 2003b. Ambient aquatic life water quality criteria for tributyltin (TBT) — final Washington: USEPA 822-R-03-031. December 2003.

[USEPA] US Environmental Protection Agency. 2003c. 2003 draft update of ambient water quality criteria for copper. Washington: USEPA-822-R-03-026. November 2003.

[USEPA] US Environmental Protection Agency. 2003d. Procedures for the derivation of equilibrium partitioning sediment benchmarks (ESBs) for the protection of benthic organisms: PAH mixtures. Washington: USEPA-600-R-02-013. November 2003.

[USEPA] US Environmental Protection Agency. 2005c. Use of biological information to better define aquatic life designated uses in state and tribal water quality standards: tiered aquatic life uses (TALU). Washington: USEPA-822-R-05-001. August 2005.

[USEPA] US Environmental Protection Agency. 2005a. Modeling framework applied to establishing an allowable frequency for exceeding aquatic life criteria — draft. Washington: December 2005.

[USEPA] US Environmental Protection Agency. 2005b. Procedures for the derivation of equilibrium partitioning sediment benchmarks (ESBs) for the protection of benthic organisms: metal mixtures (cadmium, copper, lead, nickel, silver, and zinc). Washington: USEPA-600-R-02-011. January 2005.

[USEPA] US Environmental Protection Agency. 2006. Science Policy Council peer review handbook. 3rd ed. Washington: USEPA/100/B-06-002. June 2006.

van Vlaardingen PLA, Posthumus R, Posthuma-Doodeman CJAM. 2005. Environmental risk limits for nine trace elements. Bilthoven: National Institute for Public Health and the Environment. RIVM report 601501029.

Verbruggen EMJ, Rila JP, Traas TP, Posthuma-Doodeman CJAM, Posthumus R. 2005. Environmental risk limits for several phosphate esters, with possible application as flame retardants. National Institute for Public Health and the Environment Bilthoven (NL): RIVM Report 601501024. 118 p.

Verhaar HJM, van Leeuwen CJ, Hermens JLM. 1992. Classifying environmental pollutants. 1: Structure–activity relationships for prediction of aquatic toxicity. Chemosphere 25:471–491.

Vermeirssen ELM, Burki R, Joris C, Peter A, Segner H, Suter MJ-F, Burkhardt-Holm P. 2005. Characterization of the estrogenicity of Swiss midland rivers using a recombinant yeast bioassay and plasma vitellogenin concentrations in feral brown trout. Environ Toxicol Chem 24:2226–2233.

Vos JG, Dybing E, Greim H, Ladefoged O, Lambré C, Tarazona JV, Brandt I, Vethaak AD. 2000. Health effects of endocrine disrupting chemicals on wildlife, with special reference to the European situation. Crit Rev Toxicol 30:71–133.

Wang W. 1990. Literature review on duckweed toxicity testing. Environ Res 52:7–22.

Weltje L. 2002. Bioavailability of lanthanides to freshwater organisms. Speciation, accumulation and toxicity [dissertation]. Delft: DUP Science.

Weltje L, vom Saal FS, Oehlmann J. 2005. Reproductive stimulation by low doses of xenoestrogens contrasts with the view of hormesis as an adaptive response. Human Exp Toxicol 24:431–437.

Wheeler JR, Leung KMY, Morritt D, Sorokin N, Rogers H, Toy R, Holt M, Whitehouse P, Crane M. 2002. Freshwater to saltwater toxicity extrapolation using species sensitivity distribution. Environ Toxicol Chem 21:2459–2467.

[WHO] World Health Organization. 2003. Guidelines for safe recreational water environments. Vol. 1. Coastal and fresh waters. Geneva: World Health Organization.

[WHO] World Health Organization. 2003. Guidelines for drinking water quality, 3rd ed. Geneva (Switzerland): World Health Organization.

[WHO] World Health Organization. 2006. Guidelines for drinking-water quality. 3rd ed. First addendum. Geneva: World Health Organization.

Würgler FE, Kramers PGN. 1992. Environmental effects of genotoxins (eco-genotoxicology). Mutagenesis 7:321–327.

Zabel TF, Cole S. 1999. The derivation of environmental quality standards for the protection of aquatic life in the UK. J CIWEM 13:436–440.

Zhao FJ, McGrath SP, Merrington G. 2007. Estimates of ambient background concentrations of trace metals in soils for risk assessment. Environ Pollut 148:221–229.

5 Derivation and Use of Environmental Quality and Human Health Standards for Chemical Substances in Groundwater and Soil

Graham Merrington, Sandra Boekhold,
Maria-Amparo Haro, Katja Knauer,
Kees Romijn, Norman Sawatsky, Ilse Schoeters,
Rick Stevens, and Frank Swartjes

5.1 INTRODUCTION AND SCOPE

The development of chemical standards for substances in the terrestrial compartment is a relatively new regulatory activity compared to derivation of water quality standards. Despite this, understanding of soil processes and the behavior, fate, and transport of chemicals in terrestrial and groundwater compartments is already considerable. What is still required is better understanding of how these processes influence the responses to toxicants of organisms that are of interest to the regulatory agencies that must adopt and use the standards. Development of soil standards can also be viewed as an opportunity to promote the use of robust science and understanding specific to soils rather than simply being constrained by historical paradigms and strictures considered for the aquatic compartment.

International harmonization of soil quality standards (SQSs) has been discussed in the CARACAS (Concerted Action on Risk Assessment for Contaminated Sites in the European Union, 1995 to 1998) and CLARINET (Contaminated Land Rehabilitation Network for Environmental Technologies, 1998 to 2001) concerted actions (Vegter et al. 2003), and a form of the Soil Framework Directive is still under review by member states in the European Union, so the present guidance is both timely and relevant.

Consistency, transparency, and the appropriateness of underlying methods for auditing are of great importance when deriving standards. When this is done for all media, it ensures compatibility between regulatory regimes and improves understanding by those organizations that are impacted by the standard. This chapter explores some of the key standard-setting issues for the terrestrial compartment. Guidance and reasoning are provided on the most appropriate ways forward to deliver standards that meet the range of protection goals in soils and groundwater.

For the terrestrial compartment, an SQS must be

- a numerical value related to soil
- a threshold for decisions
- related to a target (such as ecological receptors or humans) or a designated use and have a specific protection or trigger level

Due to the inherent spatial and temporal variability in soils and the resulting uncertainty of generically used standards, it is recommended that there should be few situations for which SQSs are mandatory (i.e., SQSs should not have pass-or-fail criteria in isolation from other considerations). In most cases, SQSs are a first step in a tiered approach or framework for decision making (e.g., Figure 5.1). In each step of the process, the degree of uncertainty decreases, while site specificity, and hence reliability, increases. There are few situations in which SQSs are used as compliance measures, so there is no direct need for strict pass-or-fail criteria. It should be acknowledged that a tiered system nonetheless requires 1) clear criteria associated with each specific tier, which is an issue clearly associated with initial problem formulation, and 2) clear criteria on when to pass to another tier.

5.2   STARTING POINT FOR THE DEVELOPMENT OF A TERRESTRIAL OR GROUNDWATER STANDARD

Inventories across different jurisdictions show that different types of SQSs can be classified by the goal for which they were originally intended. For example, SQSs may be established based on different soil use conditions, and these may range from environments that are quite pristine (e.g., a nature reserve) to a site that may be less than pristine (e.g., an industrial site). In addition, the different types of SQSs can be derived by a range of methods that use different amounts of data, endpoints, or protection levels, such as EC(D)10 or EC(D)50. If the goal is prevention of damage to a more pristine environment, it might make sense to use a more stringent protection level than for decisions associated with a site needing remediation. Most existing SQSs can be classified as negligible risk values, trigger values, or action values (Figure 5.2).

The SQSs can also be dependent on land use. In the derivation of human health risk-based SQSs, for example, different exposure scenarios are appropriate for different land uses. Moreover, the type of data utilized [e.g., $EC(D)_{50}$ versus $EC(D)_{10}$], the volume and quality of data required to derive an SQS, and the overall level

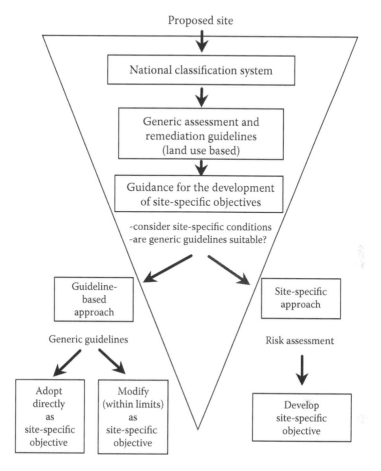

FIGURE 5.1   Canadian framework for contaminated site assessment. (Reprinted with permission from CCME 2006 © Canadian Council of Ministers of the Environment.)

of protection may differ. An example of the differentiation of SQSs for a range of possible land uses is given in Figure 5.3.

From Figure 5.3, it can be seen that protection goals can change from pristine land use in a nature reserve to industrial land use. However, this figure also incorporates the concept of a minimum level of protection (i.e., a level at which resilience of the soil ecosystem is guaranteed) that must be implemented independent of land use. Beyond this, there is the ability for policy to influence the protection level by choice. Here, the desired land use may influence decisions about the soil function that should be maintained, and a politically based land use planning decision can be incorporated into the decision-making process for soil criteria setting. For example, it may be decided at a political level that a certain land use requires greater or lesser protection due to factors other than ecological protection, such as economic considerations. Figure 5.7 gives an example of where such management and nontoxicity considerations may come into a decision-making process.

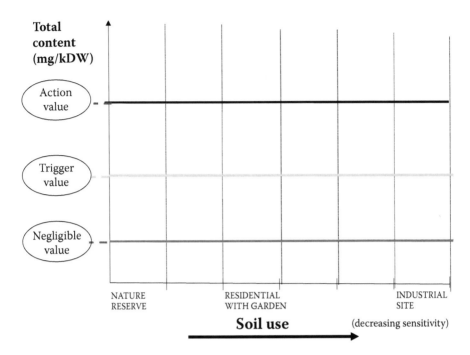

FIGURE 5.2 Different types of currently existing soil quality standards with some limited discrimination by land use.

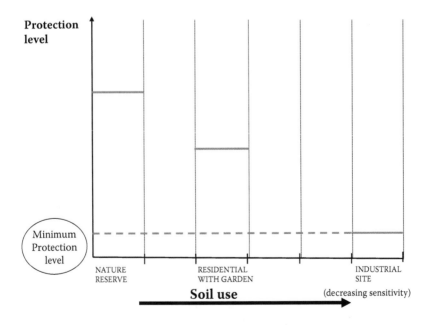

FIGURE 5.3 Different levels of protection as a function of soil use.

5.3   FURTHER CONSIDERATIONS IN SOIL QUALITY STANDARD SETTING

We recognize that social, policy, and economic considerations will most likely increasingly become part of environmental decision making. Terrestrial standards must both inform and direct the policy decisions as they apply to an area of land. Chapter 2 stresses that in the decision process there is a need to assess social and economic implications before scientific analysis and during the decision-making process. Processes, as well as the level of involvement of the SQS developer (i.e., the individuals responsible for deriving the SQS), will be different at these different stages, and this needs to be recognized during the involvement of the various stakeholders. For the process to remain transparent, clear guidelines need to be established, including guidance on the role of the SQS developer at each stage in the process.

Land use planners accept that changes in ecological function will occur as a result of decisions about land use. SQSs must be developed in a manner that not only informs the decision maker about the potential for, and nature of, effects but also directs the decision maker, when necessary, to ensure that a minimum ecological function is not completely lost from the system.

We need to articulate clearly what our standards mean and how they can be used. In addition, an SQS needs to inform the decision maker on the probability that a given decision may result in an undesirable outcome. In this way, the SQS becomes more useful in making land use planning decisions. The requirements of an ideal standard, as defined in Section 3.3.1.2, can go a long way toward creating more useful SQSs that both inform and direct the decision maker when applying science. The requirements for better information for decision makers would include the

- magnitude of impact at given concentrations
- frequency of impact at given concentrations
- duration of those impacts
- design risk

However, the concept of design risk must be amended when considering risks in the terrestrial environment. In particular, we need to recognize that there is typically a spatially dependent rather than a time-dependant risk.

Furthermore, for the terrestrial environment, criteria should be developed based on pathway–receptor relationships, including secondary pathway–receptors that may be influenced by the soil environment (e.g., groundwater, aquatic environments, secondary and tertiary consumers, or indoor air quality).

5.4   PRIORITIZATION

There are arguably 3 key drivers for prioritization of substances for development of SQSs: 1) remediation, 2) emission reductions, and 3) voluntary initiatives. The basis for any prioritization should be the consideration of some function of a substance's

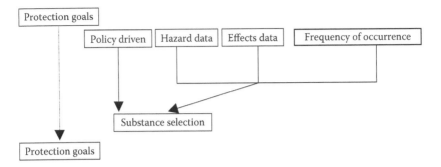

FIGURE 5.4   Some considerations in the prioritization of substances for the derivation of soil standards.

toxicity, mobility or availability, and frequency of occurrence. Figure 5.4 shows how the technical strands of prioritization may fit together in a scheme, combined with a separate policy-driven prioritization. This policy-driven approach is often for relatively new substances for which limited hazard or effect data are available but for which there is a need to provide standards. Such policy-driven substances may be selected on the basis of public or media pressure, examples of which could include methyl tert-butyl ether (MTBE); asbestos; pharmaceuticals and personal care products (PPCPs); prions; polybrominated diphenyl ethers (PBDEs); or nanomaterials (US Environmental Protection Agency [USEPA]).[1] Throughout the standard development process, including substance selection, consideration should be given to the protection goal (i.e., the aim of the original problem formulation) as well as the original standard mandate.

There are likely to be a range of considerations and drivers when selecting substances for which terrestrial standards need to be developed. There are several prioritization schemes based on hazard that can be used to rank substances, although this may be a limited approach due to the potential shortage of appropriate soil hazard data. Therefore, additional information should be used to select substances, in particular any aquatic or soil effects data on the substances considered, from actual testing or from the use of estimation approaches such as quantitative structure–activity relationships (QSARs). Further information might include the outputs from exposure models and monitoring schemes (targeted risk-based monitoring), as well as the use of production volumes to determine which substances are likely to occur in the environment.

5.5   EXPOSURE MODELS — USE IN STANDARD SETTING

Exposure models relate the concentration of a substance in soil to the potential for exposure or uptake in a human or ecological receptor. They can be described and calibrated on the basis of land use (Figure 5.3), but we recommend that there is a base set of data that would be used in calibrating minimum function in any land use. Calculated exposure differed substantially among models in a comparison of 7 exposure models (Swartjes 2002). It is therefore sensible to use more than 1 exposure model to estimate soil concentrations.

Three elements must be present and validated in any exposure model:

1) Partitioning of the substance between the soil phases (air, water, soil)
2) Transport of the substance into the contact medium (e.g., indoor air, biological tissue in plants or soil organisms, and groundwater)
3) Direct and indirect exposure to humans

A schematized example of a human health exposure model is given in Figure 5.5. This figure summarizes a generalized set of relationships between soil concentration and human exposure through several exposure pathways. It is generally recognized that the major exposure pathways concern exposure through soil ingestion, vegetable consumption and, for volatile substances, inhalation of contaminated indoor air.

For each of the exposure pathways, a relationship between soil concentration and exposure of the human receptor can be developed. Developing this relationship for each of the exposure possibilities allows the land use planner to assess how the choices for land use may influence the exposure relationships. Once the relationships between soil concentration and exposure are established, the relationships can be presented graphically, as shown in Figure 5.6. The relationships shown in this diagram are hypothetical and would ultimately depend on several chemical, physiological, and exposure factors.

For the final assessment of risk, we can then assess the exposure against a maximum permissible risk (i.e., the reference dose) to calculate a critical soil concentration associated with the exposure scenario (Figure 5.7). This information can then

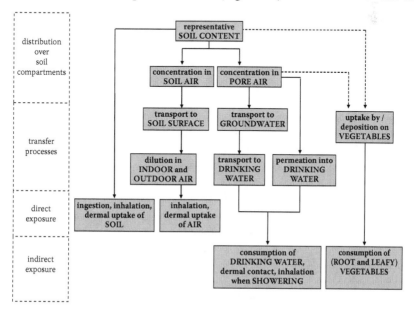

FIGURE 5.5   Schematized example of a human exposure model.

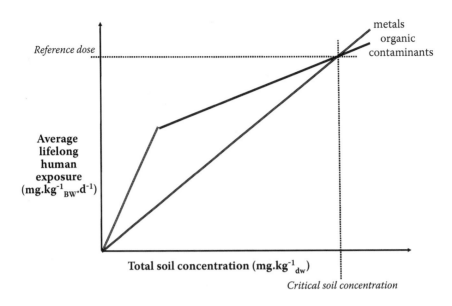

FIGURE 5.6 Hazard assessment (dose–effect characterization): hypothetical relationships between exposure and soil concentration of a chemical.

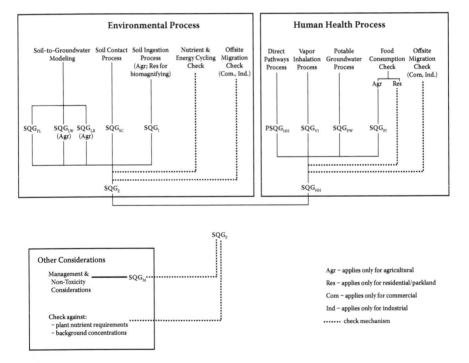

FIGURE 5.7 An example of the process that may be used in establishing criteria that are based on different exposure scenarios (CCME 2005). SQG refers to the soil quality guideline related to the specific environmental or human health process. SQG_F is the final SQG.

be used to develop a soil standard associated with the land use. Since the approach allows for assessment against different exposure scenarios, it allows for land use-dependent decisions in risk management.

When using this approach, a distinction must be made in the assessment of the exposure model between nonthreshold compounds (e.g., genotoxic carcinogens) and threshold compounds. In particular, there is a different goal-setting requirement that would be employed in calculating the reference dose, and this must be considered when applying the exposure model. In the case of carcinogenic compounds, the human protection goal is defined in additional numbers of cancer per number of inhabitants, whereas for noncarcinogens the goal is defined in terms of no occurrence of adverse vital effects.

For ecological receptors, a similar approach may be employed for secondary receptors that could be influenced by a change in the soil environment. For instance, it is possible to model the potential for soil to influence an adjacent surface water body and therefore to screen the soil criteria for impacts on an aquatic receptor. When applied, this leads to intercompartment harmonization of standards, by which soil or sediment standards pose no problems for water bodies and vice versa. In addition, it may be possible to use screening-level models to assess the potential for a bioaccumulable substance to influence a tertiary ecological receptor, usually a top predator or a protected species. In this approach, the reference dose can be borrowed from other sources (e.g., use of an aquatic criterion to determine a critical water concentration). The model is then used only to assess how the soil may influence transfer to the critical receptor. However, it should be noted that this type of procedure cannot be used for guideline development related to primary terrestrial receptors since there are no reliable models to estimate dose–response relationships for these receptors. Therefore, other techniques described in this chapter are recommended for screening against primary receptors.

5.6   ECOLOGICAL ENDPOINTS

The general aim of an ecotoxicology-related SQS is the protection of ecosystem structure and function, with the specific targets to be protected depending on the type of SQS. "Population-specific protection endpoints" relate to population performance and are usually based on tests that include growth, reproduction, and mortality. "Functional protection endpoints" relate to maintenance of key ecological services, such as litter breakdown and biochemical cycles, and include endpoints like microbial degradation and microbial endpoints related to the carbon and nitrogen cycle.

Existing EU regulatory frameworks, such as the existing substances regulation (793/93 EEC), biocides regulation, and pesticides regulation, define minimum data sets ("base sets"). These include combinations of test data, such as at least 1 data point for a plant and an invertebrate and a microbial endpoint (existing substances regulation). In addition, insects are routinely tested under the pesticides directive (91/414 EEC). Additional data are sometimes available or can be generated if necessary (e.g., more plant, invertebrate, or vertebrate species data). Endpoints addressing a specific mode of action, such as endocrine disruption, should be used as a trigger for further population studies. If sufficient data are available, they may be plotted as species sensitivity distributions (SSDs; for species- and community-level endpoints) or functional sensitivity distributions (for functional endpoints) (Posthuma et al. 2002).

Field and mesocosm data can also be used for SQS setting when available. When the goals of the study fit the goals of the SQS, these may be the most relevant data sets available, provided the study situation is similar to the problem case (e.g., for soil pH in the case of metals assessments). If the goal of the field study does not fit the protection goal of the SQS, the study results should be used as generalized supporting evidence for a decision or for validating the risk assessment steps that underpin the current quality criterion.

5.7   RELEVANCE AND RELIABILITY OF DATA

Data on fate and transport, exposure, and effects of substances on humans and other organisms in the terrestrial compartment are normally obtained from the publicly available literature. Other sources of potentially relevant data for organisms in the terrestrial compartment include industry-generated data for the process of notification, registration of pesticides and other substances, and technical reports from research institutes.

The set of data available for any given substance may vary from a simple "base set" (as discussed in Section 5.6) or perhaps less than a base set, as in the case of many existing substances, to fairly extensive data sets, as is the case for some metals, organic chemicals, and pesticides. For data-rich substances, not only are toxicity values and effects data usually available for a larger number of representative species, but also there are usually more data available for a given species for corroborating effects. The more data there are, the more reliable will be the results and the interpretation of such data. In addition, data-rich substances will potentially have been generated using a variety of conditions and different methods, representing various environmental conditions (e.g., a wide range of soil types, soil pH, percentage organic matter, and temperature).

5.7.1   SELECTION OF DATA PRIOR TO STANDARD SETTING FOR DATA-RICH SUBSTANCES

Available data can be complementary to a defined base set (e.g., representing a large number of species), but sometimes individual data points are in apparent conflict with other data (e.g., for the same species tested using 2 different methods or for different soils). "Outliers" may also be present in a data set. In both North America and the European Union, an assessment of the relevance and reliability of the available data is made prior to setting any standard.

In the United States and Canada, an assessment is made of the scientific quality of any study. A study will be classified as acceptable, supplemental, or unacceptable. Acceptable studies are those studies that were conducted according to acceptable guidelines and procedures; these studies are utilized to support fully a standard or other regulatory activity. Supplemental studies usually complement acceptable studies and may add valuable information, but supplemental studies usually cannot stand alone in setting a standard. Unacceptable studies are usually not used to help set a standard or to support any regulatory activity. In the European Union, processes for the assessment of data relevance and reliability have been published in Reach Implementation Project (RIP) 3.3 (for Registration, Evaluation, Authorization, and Restriction of Chemicals [REACH]) and in the EU technical guidance document (TGD) and the Combined Monitoring-based and Modelling-based Priority Setting (COMMPS) procedure for the Water Framework Directive.

In the COMMPS procedure, specific guidance is given on how to make an assessment of study relevance and reliability prior to setting a standard based on the available data. The data quality criteria provided in COMMPS contain the following elements:

- Endemic species are given preference over exotic species.
- The study must state the relevance of test media conditions in the environment for which standards are set (e.g., temperature, pH, percentage organic matter).
- The study should address the presence or absence of analytical supporting data (i.e., is the toxicity value based on nominal or measured concentrations?).
- There should be use of appropriate statistics to derive endpoints.
- Presence or absence of a clear dose–response effect in the data should be apparent.
- The method used should follow an established international guideline (e.g., of the Organization for Economic Cooperation and Development [OECD]).
- The study should be performed to good laboratory practice.

Additional substance-specific criteria may also be needed. Furthermore, when adopting a tiered approach, it may be worthwhile to consider that the standard data set adopted in the first tier could differ from that in higher tiers as this is a consequence of adopting tiering itself. The use of data quality selection criteria prior to setting standards is recommended to reduce variability in the data set, increase ecological relevance, and avoid setting standards based on outliers.

5.7.2   USE OF SURROGATE DATA FOR DATA-POOR SUBSTANCES

For substances that were marketed prior to introduction of specific EU legislation, the available data may not meet the base set (as defined in Section 5.6). For some substances, terrestrial data may not be available. In such cases, there is the possibility of using available aquatic data for setting soil standards. Due to uncertainty associated with the process of read-across (i.e., using aquatic data sets to set terrestrial standards), it would be more likely to use this method to set provisional values rather than establishing negligible risk or action values. Such values set by using the read-across method should be seen as tentative SQSs, not as mandatory standards, and should be used as supporting evidence in risk assessment and management.

5.8   ASSESSMENT FACTORS — EXTRAPOLATION AND SOIL QUALITY STANDARD DERIVATION

Assessment factors (AFs) are tools for dealing with the uncertainty that is implicit in any risk assessment. Three types of uncertainty should be considered (Figure 5.8):

- *Epistemological uncertainty:* This concept includes our incomplete understanding of ecological systems.
- *Methodological uncertainties:* For example, scientists have classified about 1.4 million species on our planet, which represents less than 5% of the estimated number of species currently living. Even for the least-abundant taxonomic groups, mammals and birds, the number of species reaches 4000 and 9000, respectively. The proportion of species that can be tested in toxicity studies is therefore a very small fraction of those that could be exposed.
- *Technological (including data analysis) uncertainties:* This component accounts for the uncertainty of risk assessment and focuses on the reliability of the data. For example, a toxicity test usually generates different estimates of toxicity when performed on samples from the same site by different laboratories or even by the same laboratory on different occasions. This uncertainty is often grossly underestimated.

One controversial tool to address uncertainty is the use of AFs. AFs have been discussed widely in the literature, and regulatory scientists have achieved some consensus on their use, as reflected in various TGDs. AFs will continue to be applied during the task of assessing risk and developing SQSs as long as no better tool is available.

Assessment factors for the terrestrial environment should be applied to, and focused on, the final protection goal as formulated in the problem formulation phase,

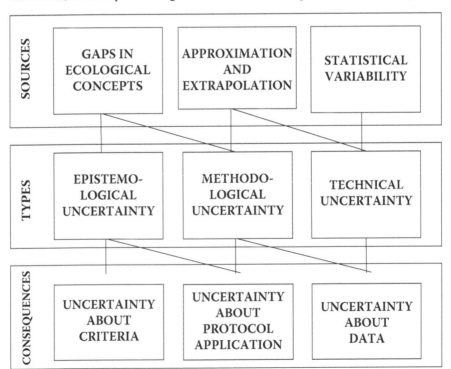

FIGURE 5.8   Types of uncertainty.

taking into account the use of the land (e.g., industrial, agriculture, residential, or natural environments). An increase in the specificity of defining a protection or risk assessment goal is often related to an increase in uncertainty, leading to the application of higher AFs or more refined extrapolation methodologies. Thus, more conservative values should be applied for natural environmental protection compared to values for industrial soils for which the protection goal is lower, as shown in Figure 5.3. In addition, more "realistic" AFs can be used if additional evidence is available from ecological and human exposure models and hazard models, especially at higher assessment tiers.

The terrestrial compartment has unique characteristics such that AFs derived using the aquatic approach may not be suitable for soil standard setting. However, evidence to support or refute the use of AFs does not currently exist. Differences between both compartments are often exposure related, and these issues should be considered during monitoring and compliance checking. In the aquatic compartment, high variations in exposure data can be found in time. This is much less of an issue in the terrestrial compartment due to the buffering capacity of the soil environment (i.e., environmental attenuation due to microbial degradation, organic matter immobilization, precipitation, and adsorption and absorption). However, the spatial variability of exposure in the soil environment tends to be greater than in the aquatic compartment, thus resulting in higher uncertainty when collecting monitoring data.

5.9   AVAILABILITY AND BIOAVAILABILITY

Total substance concentration in soil is only a gross predictor of exposure and toxicity. Indeed, total soil concentrations of substances are generally poor metrics for biological or human health effects, and the mere presence of a substance in soil may not translate into human or ecological harm. For example, if a substance in the soil is tightly bound to organic matter, it might not be available for uptake by plants or other organisms.

5.9.1   Ecological Risk Assessment

Significant variability in available toxicological effects data has been observed for some substances. Some of the variability for substances tested on the same species and for the same endpoint may be due to tests conducted in "different" media or soils. This is primarily because pH, percentage organic matter, and other soil characteristics can influence the bioavailability of substances. Hence, under certain conditions, organisms might not be exposed to the total concentration but only to a certain fraction of the total in the soil. Consideration of the bioavailable concentration in soil would ensure a better link between exposure and biological data and provide a more appropriate method for risk assessment or for quality standards derivation (International Organization for Standardization [ISO] 2008).

One method for assessing bioavailability is via pore–water concentrations. Pore–water concentrations are arguably better overall predictors of bioavailability and effects on the terrestrial ecosystem than measurements of total soil concentrations of contaminants. Peijnenburg et al. (1997) showed that for several substances risks for ecosystems and processes, as well as human risks, are related to the pore–water concentration rather than to the total soil concentration. The difficulty, however, with using soil pore–water concentration as a reliable metric for decision making is that it is both hard to measure and highly variable in temporal and spatial dimensions.

A wide range of research publications has identified various soil properties and their potential influence on substance behavior. Soil properties such as organic matter, iron, manganese, and aluminum (hydro)oxide concentrations, cation exchange capacity, and pH can all affect the bioavailability, form, and toxicity of substances.

The dependence of metal toxicity on soil characteristics has led to the development of tools to determine the bioavailable fraction of metals in soil. The terrestrial biotic ligand model (BLM) and the BLM concept model (WHAM-Model IV; Tipping et al. 1998, 2003; Di Toro et al. 2001) are the current state of the art for predicting metal bioavailability. Ma et al. (2005) have also developed a model for predicting the reduction in bioavailability of nickel and copper over time (the so-called "aging process"). These and other models can predict the fractions or available concentration of metals in soils that can be considered to be toxicologically relevant. For organic substances, the available fraction strongly depends on the organic carbon content of soil (Van den Berg et al. 1993). In summary, consideration of bioavailable concentrations ensures a better link between exposure and biological data and provides a more appropriate method for the derivation of relevant and robust quality standards and risk assessments instead of the use of total soil concentrations.

5.9.2   HUMAN HEALTH RISK ASSESSMENT

Bioavailability is assessed in a different way for human health when compared to approaches used for ecosystems.

Ingestion of vegetables and soil are 2 examples of exposure pathways for humans. For vegetables grown in soil, the accumulation of substances in vegetables consumed by humans depends not only on whether the substance will translocate into and throughout the plant but also on the available fraction of a substance in the soil. A similar approach to that described in Section 5.9.1 could be used to incorporate bioavailability assessment into the pathway of exposure through vegetable consumption.

However, for exposure through soil ingestion (e.g., a child who ingests 5 g of soil per day or for soil clinging to root crops), there is not only bioavailability of a substance in soil to consider but also the differences between intake (i.e., ingestion of substances via soil particles into the human body) and uptake (i.e., absorption of a substance into the blood and, hence, effects on a target organ). Three steps are involved in this process:

1) Bioaccessability (e.g., release of the substance from the soil particles in the stomach).
2) Transfer of a substance into the bloodstream.
3) Passage to and targeting of a specific organ, such as the liver. The difference between uptake and intake is expressed by the so-called "relative bioavailability factor" (Oomen et al. 2006).

5.9.3   GROUNDWATER

Substances applied to soil may leach and contaminate groundwater. At least 2 criteria can be used to assess the potential for substances to leach to groundwater:

1) The mobility of a substance
2) The absence of an impermeable soil layer

The absence of an impermeable layer between the soil surface, where a substance may be applied, and the groundwater increases the likelihood that a substance will leach through soil and reach the water table. For example, clay soil is considered impermeable to many substances; the absence of a clay soil layer and the presence of a sandy soil layer could increase the potential for leaching. The close proximity of a water table near the soil surface will also increase the potential for leaching.

Aging and microbial degradation of substances can also influence leaching to groundwater. Aging of substances in soil will reduce the risk of groundwater leaching. For metals, for example, adsorption and precipitation move metals from the pore–water to the solid phase on soil surfaces. Continued aging moves metals from soil surfaces to the deeper and stronger solid phase through

- surface pore diffusion
- solid-state diffusion
- occlusion of metals through precipitation of other phases
- precipitation of new metal solid phases
- occlusion in organic matter (McLaughlin 2001)

In such a case, these metals will not be subject to leaching and hence will pose no risk of groundwater contamination. For organics, microbial degradation is the most important process in reducing the concentration of a substance in soil and resulting in a lower risk of leaching to groundwater.

5.10   BACKGROUND CONCENTRATIONS

Some substances, such as metals and some organics (e.g., polycyclic aromatic hydrocarbons), occur naturally, with background concentrations in soils that can vary widely (Table 5.1). An overview of the European background concentrations of metals and other inorganic elements was provided by Salminen et al. (2006). Maps and the raw data can be found in the *Geochemical Atlas of Europe* at http://www.gsf.fi/publ/foregsatlas/.

The natural background concentration in soil of a substance is difficult to define because the ambient (measured) concentration in soil consists of both a natural pedogeochemical fraction and an anthropogenic fraction (ISO 2004). The anthropogenic fraction refers to moderate diffuse inputs into the soil or low anthropogenic pressure, not the inputs from local point sources that generally result in highly elevated concentrations. This is a pragmatic acknowledgment that it is very difficult to find soils not affected by some low-level anthropogenic additions.

TABLE 5.1

Concentrations (mg/kg) of total polycyclic aromatic hydrocarbons in soils from rural, urban, and industrial sites in the United Kingdom[a]

Percentiles	Rural	Urban	Industrial
5	0.13	0.43	0.34
50	0.72	5.39	3.49
100	168	551	1590

[a] Adapted from EA 2006.

Consideration of local or site-specific background levels of a substance is essential in any standard-setting process. Following the classical standard-setting paradigm (Figure 5.9), ecotoxicity data are gathered and are screened for their quality; finally, AFs are applied to extrapolate the data to cover uncertainties, such as interspecies variability and interlaboratory differences. These AFs are often in the order of decades (i.e., 10, 100, or 1000) and can result in an SQS very near to or within the background range of a naturally occurring substance.

The occurrence of standards within the background range of a naturally occurring substance has resulted in a situation in which standards for some compounds cannot be used in a decision-making process. The key question is how to address inherent uncertainty and provide a robust and realistic standard. If the highest uncertainty is related to background concentration, effort should be invested in improving the quality of the background data. Several options can be considered:

- Expressing the quality standard as an added value or "added SQS' approach
- Expressing the quality standard as a "total SQS' approach by one of the following options:
  - Collecting more ecotoxicity data and developing an SSD
  - Correcting the quality standard for differences in availability using speciation models (e.g., WHAM)
- Correcting the quality standard for differences in bioavailability

5.10.1 ADDED SOIL QUALITY STANDARD APPROACH

The added SQS approach assumes that only the anthropogenic added fraction of a natural element in the soil should be managed (Crommentuijn et al. 1997). The added quality standard is derived using ecotoxicity data obtained by subtracting the background values from the ecotoxicity value [EC(D)x] of the tested soils.

When checking compliance, site-specific background values are subtracted from the monitoring data, and the latter are compared with the added quality standard. In theory, the use of the added quality standard approach avoids the potential problem

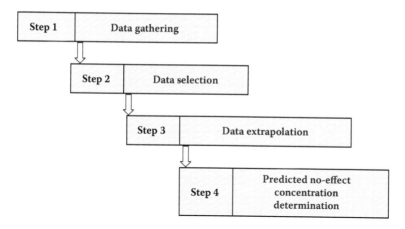

FIGURE 5.9   Stepwise approach for deriving soil quality standards.

of deriving SQS values below the background concentration. Issues to be considered when applying the added quality approach have been described by Euras (2006).

5.10.2   REFINEMENT OF THE TOTAL SOIL QUALITY STANDARD APPROACH

Several options for refinement in the derivation process of SQSs can be considered: SSD approach, correcting the quality standard for differences in chemical availability, and correcting the quality standard for differences in bioavailability.

5.10.2.1   SSD Approach

If a quality standard is derived by applying an AF of 10 or 100 on the lowest eco-toxicity value, additional ecotoxicity data could be collected sufficient to develop an SSD (OECD 1992) on which no, or a lower, AF is applied. The main underlying assumptions of the SSD approach are as follows:

- The distribution of species sensitivities follows a theoretical distribution function.
- The group of species tested in the laboratory is a random sample of this distribution.

The use of this method also implies predefining a specific level of protection (e.g., 95% of the species) when setting SQSs. This can be considered not as a real estimation of protection but as a method for using available information and establishing a common risk level. Individual data should be checked (i.e., plants, microorganisms, invertebrates, and vertebrates) and combined if appropriate. An advantage of this method is that it uses the whole sensitivity distribution of species in an ecosystem to derive an SQS instead of taking the lowest and most conservative effect endpoint. Nevertheless, collection of a random representation of species may be challenging as it is in the aquatic compartment. This approach may also not be applicable for substances with a specific mode of action. In addition, it is not known how the toxicity values, related to individual effects, translate into population consequences. Validation of the derived SQS may therefore still be required.

It is common to use the 5th percentile of the SSD to set a quality criterion from the SSD. In other cases, the lower confidence limit of the estimated HC5 (hazardous concentration to 5% of species) is used, or the HC5 is combined with an additional AF to take further uncertainties into account. When applying the AF, care should still be taken not to apply it to the background concentration. Note that the various jurisdictions that allow for the use of SSDs have adopted different "base set criteria" for SSD-based quality standard setting.

In the case of using SSD methodology, when there are data gaps, the outcome of the analysis can trigger additional research for further site-directed assessments. SSDs derived using limited data sets not fulfilling data quantity criteria defined in different jurisdictions should only be used as trigger values and "the inherent uncertainty of the output explicitly stated."

5.10.2.2   Correcting the Quality Standard for Differences in Chemical Availability

When soil parameters influencing the availability of a substance are known and models are available to predict the available fraction, SQSs can be expressed as a function of these soil properties. This approach reduces the variability of the data and increases the ecological relevance of the SQS.

This approach can be applied for substances rich in ecotoxicity data as well as for substances poor in ecotoxicity data. The prerequisite is knowledge of the soil parameter values influencing availability of the substance in the individual ecotoxicity tests. It is also important to be aware of the associated uncertainties of these data.

5.10.2.3   Correcting the Quality Standard for Differences in Bioavailability

If models are available for predicting the bioavailability of a substance in soils, these can be used to normalize the ecotoxicity data and to derive a normalized SQS related to a certain soil type. Such models have, for example, been developed for nickel and copper in soils.

Again, this approach can be applied to substances that are either rich or poor in ecotoxicity data. The prerequisites are knowledge of the soil parameter values influencing bioavailability of the substance in the individual ecotoxicity tests and applicability of the model for the species that is tested. This approach further reduces uncertainty and increases the ecological relevance of the SQS.

5.11   VERIFICATION OF THE STANDARD

It is important to verify and validate environmental quality standards for several reasons. Validation is an important aspect of cost–benefit analyses. Although it is well accepted that validation exercises can be cumbersome, there is a need for validation to establish a societal feeling of confidence and trust. Costly and important decisions should be founded on real risks and violation of real protection levels rather than on hypothetical ones. If the validation process shows a lack of validity, measures should be taken to improve the standard. Before validation, one should define the purpose of the validation because different purposes may lead to different designs of the validation methodology.

Validation applies to all standards, from no-risk standards, to trigger standards, to action standards. Standards differ in uncertainty, and this may influence the extent to which they can be validated. As SQSs are part of a tiered approach for decision making, the need for validation applies to all tiers.

5.11.1   How to Validate?

A variety of methods for validation of environmental standards can be used. The choice of the best method depends both on the standard and on the aspect that is to be validated. For example, mesocosm studies or other semifield studies may be useful for validation purposes. The use of biomarkers may also help to generate confidence in a standard. Preferably, the standard itself should be validated, but often validation for methods is lacking. In these cases, one can decide to validate the separate "building blocks" of the standard-setting process.

For example, depending on which exposure pathway is dominant, it may be important to measure either plant concentrations or indoor air concentrations since model calculations on such aspects can be uncertain.

Standards that are derived using SSDs for the soil ecosystem can in some cases be validated in the field. The overview by Posthuma et al. (2002) reported on some validation studies in which it was shown that the HC5 was lower than the no-effect concentration of studied ecosystems (i.e., in mesocosm or field conditions). An array of further studies has been published since that time. However, field studies are often difficult to interpret in terms of dose–response relationships. This difficulty in interpreting field data is sometimes due to soil heterogeneity and a highly variable soil ecosystem. Nevertheless, field soils are relevant test systems and represent a more realistic environment. Although causality may be difficult to assess, the use of pragmatic methods, derived from an expert judgment process, can improve the overall accuracy of standards.

There are some examples of validation studies in the field that show accuracy in predicting risks and effects, but even these examples are limited to specific soil and environmental conditions. In a review of toxicity information available for petroleum hydrocarbons, the authors noted that examples of validation were limited and generally restricted to agronomic environments (Bright et al. 2006). For instance, Visser (2005) attempted to examine the field response to petroleum hydrocarbons as it related to the standards developed in Canada, but field observations were restricted to 2 soil conditions and to a species choice based on agricultural applications.

Knowledge on the functioning of soil ecosystems is to a large extent lacking, and SQSs are for that reason derived using many assumptions. Therefore, there is an urgent need for more systematic validation of SQSs.

Decisions should preferably be based on as many real-life observations as possible, and additional data are frequently essential. Interpretation of data for soil quality is a challenge because of the lack of knowledge on how to distinguish natural variability in soil functions from adverse effects imposed by substances.

5.12 CONCLUSIONS

Environmental quality standards for chemical substances in groundwater and soil are important for many soil quality decisions. Examples include emission reduction measures during the marketing authorization of chemicals, soil and land use decision making and risk management, as well as soil remediation.

In providing guidance on setting SQSs (environment and human related) 3 main principles must be preserved: 1) standards need to be set in a consistent manner, 2) standards need to be transparent, and 3) standards should be audited by an external reviewer.

Social, political, and economic considerations form part of environmental decision making and are key to the process of setting standards, as discussed in Chapter 2.

Several methods have been discussed that can improve the robustness and realism of standards by reducing uncertainty. It is recommended that countries and agencies develop standards for different purposes (e.g., different land uses and different protection goals) since soils are variable by nature and have various uses and functions that may influence the protection goals. Nevertheless, a minimum level of protection or a base level is needed to guarantee resilience of the soil ecosystem, and this should be incorporated into all approaches to soil standards. Standards can be derived that indicate the presence or absence of certain data. For example, if certain ecotoxicological effects data are missing, a standard may still be established, with considerable certainty, for a minimum level of protection. However, in many cases more information may be necessary to reach a conclusion about the level of risk of a substance in soil to help make decisions about risk-based land management, including soil remediation.

Exposure models are a key tool in deriving SQSs for human health and environmental quality. These models should include all relevant soil processes and descriptions of human and animal behavior. Exposure models can be used to assess the likely spatial extent and severity of standard exceedance and provide input to prioritization of substances for which soil standards may be developed.

For many substances, reliable and relevant soil-to-biota transfer and fate data for the terrestrial environment are lacking. For data-poor substances, aquatic data may be used to derive terrestrial standards, but great care should be taken because use of surrogate aquatic data increases uncertainty.

The SQS must take into account the background concentrations of naturally occurring substances.

As the aim of an SQS is the protection of ecosystem function and structure or appropriate risk management and remediation, the most relevant toxicity endpoints for setting SQSs are quantified on the basis of toxicant effects on endpoints, such as growth, mortality, and reproduction. Biomarkers and endpoints related to the mode of action, such as endocrine disruption, should be used to trigger further studies on population effects. Field studies can be used to set an SQS but only if the goals of the original study match the protection goals of the SQS. If this is not the case, field studies should only be used as supporting evidence and for validation of the risk assessment models underlying the quality standards.

As only a small fraction of a substance in the soil is likely to be available to organisms, total concentrations of substances in soils are generally poor predictors of toxicity. The robustness of a quality standard can be increased considerably by accounting for the bioavailability of the substance. Bioavailable concentrations are influenced by the physical and chemical properties of a substance, soil characteristics, weather, and land use. Further international efforts are necessary to gain understanding of the main regulating processes of availability and bioavailability to improve the prediction of risks for substances in soil and to derive more accurate SQSs.

NOTE

1. Personal communication with R. Stevens, US Environmental Protection Agency, 2006. Emerging contaminants evolving policy in the Office of Water.

REFERENCES

Bright DA, Sanborn M, Sawatsky N. 2006. Relative sensitivity of different soil-associated flora and fauna to petroleum hydrocarbon releases: current state of the knowledge and implications for environmental protection goals [draft]. Edmonton (Alberta): Alberta Environment.

Carlon C, editor. Forthcoming. Derivation methods of soil screening values in Europe. A review of national procedures towards harmonisation opportunities. Ispra, (It): European Commission Joint Research Centre. p. 32.

[CCME] Canadian Council for Ministers of the Environment. 2004. Canadian environmental quality guidelines. Winnipeg (Manitoba): Available from: <http://www.ec.gc.ca/ceqg-rcqe/english/download/default.cfm>. Accessed 30 June 2009.

[CCME] Canadian Council for Ministers of the Environment. 2005. A protocol for the derivation of environmental and human health soil quality guidelines. Available from: <http://www.ccme.ca/assets/pdf/sg_protocol_1332_e.pdf>. p. 215. Accessed 30 June 2009.

Crommentuijn T, Polder MD, van de Plassche EJ. 1997. Maximum permissible concentrations and negligible concentrations for metals, taking background concentrations into account. RIVM report no. 601501001.

Di Toro DM, Allen HE, Bergman HL, Meyer JS, Paquin PR, Santore RC. 2001. Biotic ligand model of the acute toxicity of metals. 1. Technical basis. Environ Toxicol Chem 20:2383–2396.

[EA] Environment Agency. 2006. UK soil and herbage survey. Report no. 9. Bristol (UK).

Euras. 2006. Fact sheet 3: risk characterization, general aspects. MERAG program-building block, risk assessment, Annex 1, added versus total risk approach and its use in risk assessment and/or environmental quality setting.

[ISO] International Organization for Standardization. 2004. Soil quality: guidance on the determination of background values. Geneva: ISO CD 19258.

[ISO] International Organization for Standardization. 2008. Soil quality: requirements and guidance for the selection and application of methods for the assessment of bioavailability of contaminants in soil and soil materials. Geneva: ISO 17402:2008.

Ma Y, Lombi E, Nolan AL, McLaughlin MJ. 2005. Short term natural attenuation of copper in soils: effects of time, temperature and soil characteristics. Environ Toxicol Chem 25:652–658.

McLaughlin MJ. 2001. Ageing of metals in soils changes bioavailability. Fact sheet 4 on environmental risk assessment. ICME (International Council on Metals and the Environment). 6 p.

[OECD] Organization for Economic Cooperation and Development. 1992. Report of the OECD workshop on the extrapolation of laboratory aquatic toxicity data on the real environment. Paris: OECD environment monographs no. 59.

Oomen AG, Brandon EFA, Swartjes FA, Sips AJAM. 2006. How can information on oral bioavailability improve human health risk assessment for lead-contaminated soils? Implementation and scientific basis. Bilthoven: RIVM. RIVM report 711701042. Available from: <http://rivm.openrepository.com/rivm/bitstream/10029/7433/1/711701 042.pdf>.

Peijnenburg WJD, Posthuma L, Eijsackers HJP, Allen HE. 1997. A conceptual framework for implementation of bioavailability of metals for environmental management purposes. Ecotoxicol Environ Saf 16:163–172.

Posthuma L, Suter GW, Traas TP. 2002. Species sensitivity distributions in ecotoxicology. Pensacola (FL): CRC Press.

Salminen R, chief editor, Batista MJ, Bidovec M, Demetriades A, De Vivo B, De Vos W, Duris M, Gilucis A, Gregorauskiene V, Halamic J, Heitzmann P, Lima A, Jordan G, Klaver G, Klein P, Lis J, Locutura J, Marsina K, Mazreku A, O'Connor PJ, Olsson SÅ, Ottesen R-T, Petersell V, Plant JA, Reeder S, Salpeteur I, Sandström H, Siewers U, Steenfelt A, Tarvainen T. 2006. Geochemical atlas of Europe. 1. Background information, methodology and maps. Vienna: GTK, FOREGS.

Swartjes FA. 2002. Variation in calculated human exposure: comparison of calculations with seven European exposure models.Bilthoven (NL): RIVM. RIVM report 711701030.

Tipping E, Lofts S, Lawlor AJ. 1998. Modelling the chemical speciation of trace metals in the surface waters of the Humber system. Sci Total Environ 210:63–77.

Tipping E, Rieuwerts J, Pan G, Ashmore MR, Lofts S, Hill MTR, Farago ME, Thornton I. 2003. The solid-solution partitioning of heavy metals (Cu, Zn, Cd, Pb) in upland soils of England and Wales. Environ Pollut 125:213–225.

Van den Berg R, Denneman CAJ, Roels JM. 1993. Risk assessment of contaminated soil proposals for adjusted, toxicologically based Dutch soil clean-up criteria. In: Arendt et al., editors. Contaminated soil 93. Dordrecht (NL): Kluwer Academic.

Vegter J, Lowe J, Kasamas H, editors. 2003. Sustainable management of contaminated land: an overview. (AU): Umweltbundesamt GmbH. p. 115.

Visser S. 2005. Toxicity of petroleum hydrocarbons to soil organisms and the effects on soil quality, phase 3: long-term field studies. Report prepared for Petroleum Technology Alliance Canada (PTAC). Calgory (Alberta, CN). www.ptac.org.

6 Workshop Conclusions and Recommendations

*Mark Crane, Dawn Maycock,
and Graham Merrington*

The SETAC (Society of Environmental Toxicology and Chemistry) Technical Workshop on the Derivation and Use of Environmental Quality and Human Health Standards for Chemical Substances in Water and Soil produced the following main conclusions and recommendations:

1) Much standard setting to date has been developed in a piecemeal fashion with little consistency between schemes in the levels of protection sought, the selection of chemicals for which standards may be needed, the methods used to derive them, or the methods used to monitor compliance. These differences can lead to the implementation of substantially different values from the same empirical data, which must mean that their application is either over- or underprecautionary in at least some situations.

2) It is appropriate to have different types of environmental quality standards (EQSs, often referred to as soil quality standards [SQSs] for soils) to protect the environment and human health, but there are many more ways to set and implement a standard than legally binding limits that are introduced through "direct" regulation. However, even the "softer" approaches entail the use of a numerical standard that must have a sound basis in science.

3) Stakeholders (especially policy makers) must agree on the following when specifying a standard:
   a. There is a need for a standard.
   b. The technical ability exists to deliver it.
   c. There are no insurmountable constraints.
   d. The stakeholders have been identified.
   e. There is an agreed mandate from which the process can begin.

4) Early consideration of implementation options during the specification stage will ensure that resources are not wasted on deriving standards that cannot be used and will help to focus effort on the best solutions.

5) It is possible to place most standards into 2 main groups:
   a. Type A values are standards that if met are not anticipated to result in adverse environmental or human health effects. Exceedance of these standards does not necessarily lead to adverse environmental

consequences because they tend by design to be conservative and precautionary.

b. The second type of values, Type B, is standards that if exceeded are anticipated to result in adverse environmental effects.

6) There should be flexibility in the way a standard is applied to reflect the levels of uncertainty associated with the standard and the consequences of failure. If the consequences of failure are controversial or expensive, it makes sense to demand that failure is demonstrated with high confidence before such action is imposed on the unwilling. This gives the statistical benefit of the doubt to the polluter. If less dramatic action is to be imposed (e.g., a requirement for further studies), we might act on less confidence of failure. Costly action must not be imposed solely because a regulator has chosen not to monitor adequately.

7) Five points that robust or "ideal" standards should reflect for a chemical in flowing waters where compliance is assessed by routine periodic monitoring are as follows:

a. A measurable limit value (e.g., a concentration of 10 µg L^{-1}).

b. A summary statistic, such as how often the limit may be exceeded (e.g., 5% of the time or on 5% of the monitoring events). This point excludes the absolute limit because compliance and planning for such standards can only be done by defining the absolute limit as a particular percentile. Instead of an absolute limit, statistics like the mean or percentiles are required.

c. The period of time over which this statistic is calculated, such as a calendar year.

d. The definition of the design risk, which is the proportion of time periods for which failure to meet the standard is planned as accepted, such as 1 in 20 calendar years. This will be relevant in designing the action needed to correct failure. In other words, when this action is complete, how often is it acceptable for failure to be recorded?

e. The statistical confidence with which noncompliance is to be demonstrated before acting on failure.

8) For air, soils, sediments, groundwater, lakes, and marine waters, the first point for the ideal standard can be generalized to

a. A limit that refers to spatial considerations, such as a concentration of 10 mg kg^{-1}. This might be a spatial average (or an estimate of this) over a specified depth, area, or volume. Alternatively, it might be a proportion of such, for example, the concentration exceeded by 10% of the area.

b. A summary statistic for these spatial concentrations with respect to time or space. For example, how often this limit may be exceeded — say, 5% of the time — or to establish there is no spatial trend. This might be a trivial step in cases of no temporal variation or true spatial heterogeneity.

9) There is no consensus about the minimum amount of data required for deriving EQSs, but the use of very small data sets (e.g., toxicity data on 1 alga, 1 crustacean, and 1 fish species) is likely to result in unreliable predicted no-effect concentrations (PNECs) from which EQS limit values

are derived. Toxicity data from which EQSs are derived must have been validated according to an agreed procedure, especially if they derive from nonstandard tests. Furthermore, it is important to ensure that the derivation procedures themselves, and all decisions resulting from evidence-based judgment, are properly validated, recorded, and independently reviewed.

10) Substances generally require standards protective against both short- and long-term exposure, especially in aquatic systems, and often are expressed as a not-to-be-exceeded and a long-term concentration (e.g., maximum acceptable concentrations [MAC-EQSs] and annual average concentrations [AA-EQSs]). While acute toxicity data are needed for deriving MAC-EQSs, it is preferable to use chronic toxicity data for deriving AA-EQSs.

11) While EQSs for aquatic and terrestrial ecosystem protection can be derived solely using assessment factors applied to data from test species, it is considered more reliable to use species sensitivity distributions (SSDs) based on larger chronic data sets (if available), and the use of good-quality model ecosystem data (again, if available) is probably the most reliable approach for deriving long-term EQSs, particularly if exposure in the tests was maintained. Mesocosms in which exposure was transient may also be used for deriving short-term EQSs.

12) Assessment factors, whose size should be designed to reflect expected uncertainties in the data, should generally be used when deriving EQS values from either laboratory or model ecosystem data. It may also be appropriate to apply small assessment factors to the outcome of SSDs.

13) There is no "systematic" bias in chemical sensitivity between freshwater and marine species, but use of freshwater data to support the derivation of marine EQSs should be conducted with caution on a case-by-case basis. Marine EQSs based on "read-across" from freshwater data should be regarded as tentative rather than definitive. Furthermore, the somewhat higher biodiversity in the marine ecosystem as a whole should not automatically result in the use of higher assessment factors when deriving marine EQSs.

14) The aquatic food chain to humans and wildlife should be protected by calculating expected tissue concentrations that are predicted to result from ingestion of aquatic organisms (based on measured bioconcentration factors, possible biomagnification, and frequency of ingestion) and comparing these with toxicity data (acceptable daily intake) for birds, mammals, and so on. In the case of humans, this process must be adjusted to account for losses during food preparation. EQSs may be derived from internationally agreed acceptable daily intake values and applied to raw waters for the protection of humans and animals from their ingestion, but it is often sufficient to apply these values to treated water if the treatment technology is adequate. Drinking water quality is best protected through the use of holistic Water Safety Plans that include source water protection as well as control hazards and risks throughout the water supply system. EQSs for protecting recreational uses of water can be aimed at synthetic chemicals (especially

in acute spill situations) but usually are only relevant for microorganisms and algal toxins.

15) Soil Quality Standards are developed for different purposes (e.g., different land uses and different protection goals) since soils are variable by nature and have various uses and functions that may influence the protection goals. Nevertheless, a minimum level of protection or a base level is needed to guarantee resilience of the soil ecosystem, and this should be incorporated into all approaches to soil standards. Standards can be derived that indicate the presence or absence of certain data.

16) Exposure models are a key tool in deriving SQSs for human health and environmental quality. These models should include all relevant soil processes and descriptions of human and animal behavior. Exposure models can be used to assess the likely spatial extent and severity of standard exceedance and to provide input to prioritization of substances for which soil standards may be developed.

17) For many substances, reliable and relevant soil-to-biota transfer and fate data for the terrestrial environment are lacking. For data-poor substances, aquatic data may be used to derive terrestrial standards, but great care should be taken because use of surrogate aquatic data increases uncertainty. Methods for calculating sediment and soil EQSs from water-phase EQSs (e.g., equilibrium partitioning), for predicting toxicity based on biomarker responses, or for predicting toxicity based on chemical structure should be treated with caution and should generally result in tentative rather than definitive EQS values.

18) The geographical scope of a standard may need to be given consideration, especially when spatial variations exist in factors affecting the toxicity limit value of a standard. This may be through a desire to protect particular species, which may be rare or of particular ecological, economic, or recreational importance to a local area or region, or through the need to take into consideration factors that have a significant effect on the exposure of a substance, such as the bioavailability of some trace metals.

19) A distinction sometimes needs to be made between the natural background level of a substance, which arises purely as a result of natural processes, and the ambient background level, which is the concentration measurable in the environment at a "pristine" site. In practice, a pristine site is often considered to be one that does not receive any direct inputs from local sources, although it needs to be accepted that for many substances there may be an appreciable input from diffuse atmospheric sources. Although the scientific basis for taking background concentrations into account may be questioned (e.g., the origin of a substance cannot be distinguished by the affected organism or by analysis), there may be instances when approaches such as "added risk" are considered for use as pragmatic risk management tools. There is no consensus on the appropriateness of using background concentrations as the starting point for assessing EQS compliance. The jurisdictions represented at the workshop have a variety of approaches to addressing "natural substances" (e.g., metals). In the European Union, aquatic EQSs for natural

substances such as metals may need to be modified locally to take account of background concentrations in waters and sediments, provided that the environmental concentration exceeds the EQS. Action should only be taken if the measured environmental concentration exceeds the background concentration for the area in question when added to the EQS concentration. In the United States, the EQS derivation process for metals (and many other substances) is designed to be site specific and thus addresses the unique water, sediment, and species of each at the time of derivation. An "added risk approach" would be duplicative.

20) The robustness of a quality standard can be increased considerably by accounting for the bioavailability of the substance in an approach that is more accurate than use of added risk. The validation and implementation of international standards for such approaches is desirable. Bioavailable concentrations are influenced by the physical and chemical properties of a substance, soil characteristics, weather, and land use. Further efforts are necessary to gain understanding of the main regulating processes of availability and bioavailability to improve the prediction of risks for substances in soil and to derive more accurate SQS.

21) The success and accuracy with which MAC- and AA-EQSs are used will depend crucially on the design of the monitoring programs that provide the data on chemical concentrations in waters, sediments, and soils, which in turn should depend on the properties of the chemical and its discharge or distribution pattern. Exceedance of an EQS may result in immediate regulatory action but should often lead to further chemical or biological investigations. EQSs should not generally be used in isolation but as one of several lines of evidence about the likely impact of a chemical.

22) There may be occasions when a standard is set at a concentration below current analytical limits of detection (LODs) or limits of quantification (LOQs). This could be because high uncertainty leads to the application of large assessment factors to toxicity data to derive a standard or because analytical techniques for a particular environmental matrix have higher LODs or LOQs than those available for the medium in which the standard was derived (e.g., sewage effluent versus laboratory water). An inability to measure concentrations of a chemical at the standard does not necessarily render the standard totally useless. For example, a water quality standard set in a receiving watercourse may be below the LOD or LOQ, but measurement of concentrations from an effluent may be above these limits. Appropriate modeling may allow good estimation of whether the standard in the water course has been exceeded.

23) An assessment of whether a particular standard has met its original protection objectives should always be an integral part of the standard-setting process. This should be conducted along with a review of the standard should there be information that can be used to update it and increase overall confidence in its application. A stringent standard may be more likely to meet its protection goals, but it is also important to consider the social and economic aspects surrounding its implementation because it is also important that

standards do not place unnecessary burdens beyond what may be required to achieve objectives.

24) Substances that are carcinogenic, mutagenic, or reproductively toxic (i.e., CMRs), for example, some endocrine disrupters, may pose special problems for derivation of EQSs (e.g., lack of internationally agreed tests in some cases; difficulties with prediction of "safe" concentrations), but use of special tests for these properties is only justified for a small subset of chemicals that meet clear criteria. Furthermore, EQSs for these substances should not be derived directly from in vitro data or from biomarkers of exposure but from in vivo tests alone.

25) Many countries conduct EQS derivations on the same substance, often by somewhat different methods, which is highly inefficient and wasteful of resources. To improve this situation, we recommend that the advantages and disadvantages of international sharing of EQS data and derivation strategies should be examined. Drivers for this include the following considerations:

    a. Even after more than 30 years of work, most countries have developed fewer than 50 EQSs for the aquatic environment.

    b. Development of a single EQS usually requires at least 2 to 3 years and can cost US$50K to US$150K or more, depending on data availability, levels of uncertainty that must be resolved, and any economic or social controversies about the substance.

    c. Most countries have similar priority substances.

    d. Duplication of work is wasteful; there is great potential for collaboration.

    e. Most EQS derivation procedures are fairly similar, so there is potential for international harmonization.

    f. Pollution often straddles national borders.

    g. Industry and trade are multinational; that is, sources of pollution are international.

26) As a first step, it would be useful if access to data on specific substances could be made simpler and a consensus reached about which data are sufficiently reliable for EQS derivation. A long-term goal would be the derivation of internationally agreed, science-based benchmarks that could be the basis (starting point) for national EQS (or criteria, guideline, objective) derivation programs. This recognizes that different jurisdictions will have different political priorities, but that the resulting standards should still have a common scientific basis.

Index

Other Titles from the Society of Environmental Toxicology and Chemistry (SETAC)

Freshwater Bivalve Ecotoxicology
Farris, Van Hassel, editors
2006

Estrogens and Xenoestrogens in the Aquatic Environment:
An Integrated Approach for Field Monitoring and Effect Assessment
Vethaak, Schrap, de Voogt, editors
2006

Assessing the Hazard of Metals and Inorganic Metal Substances
in Aquatic and Terrestrial Systems
Adams, Chapman, editors
2006

Perchlorate Ecotoxicology
Kendall, Smith, editors
2006

Natural Attenuation of Trace Element Availability in Soils
Hamon, McLaughlin, Stevens, editors
2006

Mercury Cycling in a Wetland-Dominated Ecosystem:
A Multidisciplinary Study
O'Driscoll, Rencz, Lean
2005

Atrazine in North American Surface Waters:
A Probabilistic Aquatic Ecological Risk Assessment
Giddings, editor
2005

Effects of Pesticides in the Field
Liess, Brown, Dohmen, Duquesne, Hart, Heimbach, Kreuger, Lagadic,
Maund, Reinert, Streloke, Tarazona
2005

Human Pharmaceuticals: Assessing the Impacts on Aquatic Ecosystems
Williams, editor
2005

SETAC

A Professional Society for Environmental Scientists and Engineers and Related Disciplines Concerned with Environmental Quality

The Society of Environmental Toxicology and Chemistry (SETAC), with offices currently in North America and Europe, is a nonprofit, professional society established to provide a forum for individuals and institutions engaged in the study of environmental problems, management and regulation of natural resources, education, research and development, and manufacturing and distribution.

Specific goals of the society are

- Promote research, education, and training in the environmental sciences.
- Promote the systematic application of all relevant scientific disciplines to the evaluation of chemical hazards.
- Participate in the scientific interpretation of issues concerned with hazard assessment and risk analysis.
- Support the development of ecologically acceptable practices and principles.
- Provide a forum (meetings and publications) for communication among professionals in government, business, academia, and other segments of society involved in the use, protection, and management of our environment.

These goals are pursued through the conduct of numerous activities, which include:

- Hold annual meetings with study and workshop sessions, platform and poster papers, and achievement and merit awards.
- Sponsor a monthly scientific journal, a newsletter, and special technical publications.
- Provide funds for education and training through the SETAC Scholarship/Fellowship Program.
- Organize and sponsor chapters to provide a forum for the presentation of scientific data and for the interchange and study of information about local concerns.
- Provide advice and counsel to technical and nontechnical persons through a number of standing and ad hoc committees.

SETAC membership currently is composed of more than 5000 individuals from government, academia, business, and public-interest groups with technical backgrounds in chemistry, toxicology, biology, ecology, atmospheric sciences, health sciences, earth sciences, and engineering.

If you have training in these or related disciplines and are engaged in the study, use, or management of environmental resources, SETAC can fulfill your professional affiliation needs.

All members receive a newsletter highlighting environmental topics and SETAC activities and reduced fees for the Annual Meeting and SETAC special publications.

All members except Students and Senior Active Members receive monthly issues of Environmental Toxicology and Chemistry (ET&C) and Integrated Environmental Assessment and Management (IEAM), peer-reviewed journals of the Society. Student and Senior Active Members may subscribe to the journal. Members may hold office and, with the Emeritus Members, constitute the voting membership.

If you desire further information, contact the appropriate SETAC Office.

1010 North 12th Avenue
Pensacola, Florida 32501-3367 USA
T 850 469 1500   F 850 469 9778
E setac@setac.org

Avenue de la Toison d'Or 67
B-1060 Brussels, Belgium
T 32 2 772 72 81   F 32 2 770 53 86
E setac@setaceu.org

www.setac.org
Environmental Quality Through Science®